Math to Know

Teacher's Resource Book

Mary C. Cavanagh

GReaT S*uRCe®
EDUCATION GROUP
A Houghton Mifflin Company
New Ways to Know™

Credits:

Design and Production: Taurins Design Associates

Illustration Credits:

Robot Characters: Terry Taylor

Creative Art: Joe Spooner *page 9*, Scott Ritchie *pages 11, 12, 23, 26, 30, 32, 37, 39, 61, 69, 74, 77, 85, 89, 91, 99, 101, 105, 118, 121, 136, 138, 139, 140, 150, 151, 152, 153, 157, 161, 165, 178, 181, 185, 188, 190, 191, 200, 203, 205, 210, 213, 215, 216, 218, 220.*

Technical Art: Taurins Design Associates

Printed in the United States of America

Great Source® and *New Ways to Know*® are registered trademarks of Houghton Mifflin Company.

International Standard Book Number -13: 978-0-669-48075-7

International Standard Book Number -10: 0-669-48075-4

10 11 12 13 1412 15 14

4500314076

Visit our website: http://www.greatsource.com/

Math to Know is a reference handbook for students, teachers, and parents. It provides concise explanations and examples that are written on the student level. *Math to Know*, with instructional support from teachers, is a tool that can empower students to become more responsible for their own learning, reviewing, relearning, research, and extended thinking.

TEACHER USES

Math to Know is organized by topics, not by chapters. Some students may need only a quick review of basic fraction concepts, but need more instruction in computing with fractions. *Math to Know* brings all those skills together in one section.

Vocabulary is an area of mathematics that is often overlooked or not emphasized. *Math to Know* provides an avenue for a stronger approach to terms and their applications since it contains definitions and explanations that are easy to communicate to students. The use of vocabulary in instruction also provides ways for students who have difficulty with computation to excel through their understanding of verbal relationships and meanings.

STUDENT USES

Math to Know provides a ready reference for students whose notes are unavailable, incomplete, or indecipherable when they are doing homework. There are times when textbook explanations need further clarification. *Math to Know* gives clear explanations that allow students to understand a difficult topic more fully. The pictures, charts, and simple explanations fill gaps in learning that a student may have, but is unlikely to voice.

Student research is an important part of education. Students can easily access information in *Math to Know* through various points of entry. They can look up a term in the glossary, found in the Yellow Pages section. The terms are cross-referenced to pages in the main section of the handbook. Each section is color coded with a key on the back cover. The use of mathematically correct vocabulary along with the examples helps students connect words with symbols. The glossary provides a comprehensive resource of math terms.

PARENT USES

Math to Know is a concise handbook of math topics that parents can readily use. Math textbooks usually have detailed explanations of single skills. *Math to Know* presents those single skills in a larger context and connects concepts to previous learning. Many parents may find this a useful mechanism for helping them recall mathematical steps they may have forgotten.

The wide variety of diagrams and clear explanations in *Math to Know* gives parents a great resource to help with homework. Equipping parents to help improve their child's learning is a benefit that reaches far beyond the classroom.

The organization of this book parallels the sections of *Math to Know*. You may first want to correlate the chapters of your textbook to the sections of *Math to Know*. This will enable a ready reference to the practice and activities in this book. Each section may include more than one chapter of your text.

Math to Know Teacher's Resource Book provides practice and test preparation pages for each sub-section of *Math to Know*. These pages are followed by an answer page with answers to both the practice and test questions. Each section concludes with an application activity, designed to connect topics that relate to more than one area of mathematics.

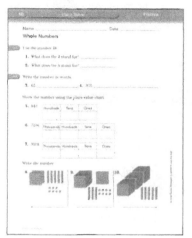

The practice pages are copymasters that follow the organization of each sub-section of *Math to Know*. Each problem is cross-referenced to show students which page(s) in *Math to Know* will provide help, if needed. This feature also allows you to identify specific concepts that students may need help to understand. The practice pages can be assigned by sub-section in their entirety or by question number to accompany material recently covered.

The test preparation pages are copymasters that present the concepts of each sub-section of *Math to Know* in a variety of standardized state test formats. This enables students to become familiar with question formats they may encounter when taking standardized tests. Students who have had experience with these formats often achieve higher test scores than students who have had no prior encounters with the types of questions used in standardized tests.

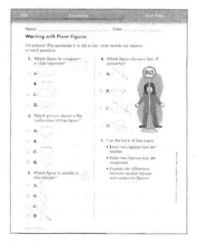

The application activities at the end of each section are designed to provide examples of mathematics as it relates to real-world situations. Some activities relate directly to daily tasks, while some involve using math manipulatives to reinforce concepts. Each application includes extension activities on the teacher page so that students may continue studies on the topic. Also in this book is a rubric that you may use for grading many of the activities.

At Your Fingertips

OBJECTIVE
- Explore the *Math to Know* handbook to determine the format of the book

MATERIALS
- *Math to Know* handbook

TIME
- 30 minutes

TEACHER NOTES
- This activity leads students on a tour through the *Math to Know* handbook and enables them to become more familiar with the organization and features of the book. It also introduces them to the structure of the application activities of this book.

- Have students correlate chapters of their textbook to the sections of *Math to Know*. Often more than one chapter may relate to a section of the handbook.

EXTENSIONS
- Have students find four mathematical terms in the Glossary that are new to them. Have them write an explanation for each term in their own words, so they could explain the new terms to a friend.

- Have students work in pairs. One student writes down an exercise which is worked out in the handbook, for example, $10 - 2 + 4 \times 3 \div 2 = \blacksquare$. The other student looks through the handbook until he or she locates the page that that exercise is on. In this case, the exercise is on page 253.

ANSWERS
1. Each section has its own color.
2. page 208
3. page 258, page 258, page 258
4. the Glossary
5. 3 ways
6. page 90   Multiples
   page 210   Fraction Concepts
   page 220   Equivalent Fractions
   page 221   Writing a Fraction in Simplest Form

Name _____ Date _____

This spot will list references to *Math to Know* to help you find information you can use to solve the problems on the activity pages. This activity will help you discover how the Math to Know handbook is organized.

At Your Fingertips

1. How can you tell the section on Place Value from the section on Basic Operations?

   _____

2. Find the section on Fractions. What page does it start on? _____

3. Pages x–xi, "How to Use This Book," show three ways to find information about math topics. Look up *Ordered Pairs* in each of the three sections and write the page number(s) you find.

   Index _____ Glossary _____ Table of Contents _____

4. Suppose you want to find definitions for the terms *prime number* and *composite number*. What is the quickest place to look in *Math to Know* to find the

   definitions? _____

5. *Math to Know* often shows more than one way to find an answer. Look at pages 180–181. How many

   different ways are shown to find 15×28? _____

6. For many topics in *Math to Know* there are More Help messages. These messages refer you to other topics that might help you. Look at the section on Decimals and Fractions on page 30. Write the More Help page numbers that are given. Go to each page and write the topic you find there.

   Page (s)   Topic

   _____   _____

   _____   _____

   _____   _____

   _____   _____

Name _____ Date _____

	Gold	Silver	Bronze	Copper
Comprehension	Specific facts and relationships are identified and well-defined.	Most facts and relationships are defined.	Some facts are identified but relationships are missing.	No facts or relationships stated.
Application and Analysis	A strong plan is developed and executed correctly.	A plan is developed and implemented with some computational errors.	Some organized computation toward a weak plan.	Random computation with little relation to problem. No plan is present.
Mechanics	Appropriate and correct computation, mathematical representation, and graphics.	Appropriate computation that may be incorrect. Mathematical representation and graphics are present.	Computation is wrong and leads to further mistakes.	Computation is random. Mathematical representations are non-existent.
Presentation	Strong and succinct communication of results.	Strong communication of results. Justification for method may be weak.	Communication of results is present, but lacks any justification.	No results are communicated. No justification is to be found. A correct answer may have appeared.
Aesthetics	Exceptional. Attractive. Encourages attention. All requirements exceeded.	Neat and orderly. Requirements met.	Messy and disorganized. Some requirements missing.	Illegible and random information. Most requirements missing.

Name _____ Date _____

	Gold	Silver	Bronze	Copper
Do I know what to do?	I used the right numbers and the right plan.	I used some right numbers and an OK plan.	I used some right numbers but I need a plan.	I haven't got a clue.
Can I do it?	I got it all done and can defend my answer(s).	I got it all done and most of it is right.	I got some of it done and I think it's right.	I got some answers but I'm not sure they're right.
Is it right?	There are no mistakes. I think my pictures are great.	Oops! I made a couple of careless errors.	I made mistakes that messed up the other answers.	I can't tell if it's right.
Is my work clear?	Everyone could understand my work.	I may need to explain some of my work.	I'll definitely need to explain my work.	I hope no one asks me about my work.
How does it look?	Wow! Maybe I can help somebody else next time!	It could have been neater.	This looks really unorganized.	I should have asked for help.

Name _____ Date _____

Whole Numbers

2–4 Use the number 48.

    1. What does the 4 stand for? _____

    2. What does the 8 stand for? _____

4–7 Write the number in words.

    3. 62 _____   **4.** 975 _____

Show the number using the place-value chart.

5. 843

Hundreds	Tens	Ones

6. 7256

Thousands	Hundreds	Tens	Ones

7. 3018

Thousands	Hundreds	Tens	Ones

Write the number.

8.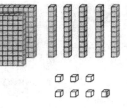

    _____

9.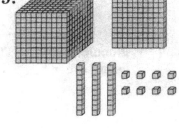

    _____

10.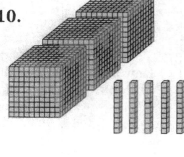

    _____

Name _____ Date _____

Write the value of the underlined digit.   8–11

11. 1,603,500 _____

12. 32,790,418 _____

13. 95,034,820 _____

Compare the numbers. Use < or >.   12–13

14. 38 ◯ 52   15. 461 ◯ 437   16. 597 ◯ 1304

17. Order the numbers from least to greatest: 73, 29, 28   14–15

_____

18. Order the numbers from greatest to least: 2561, 957, 2378

_____

Write the word name for each ordinal number.   16

19. 1st _____   20. 4th _____

21. 25th _____   22. 32nd _____

23. Who is the fourth person in line? _____

24. What position is Fred in? Write the word name

for the ordinal number. _____

Name _____ Date _____

Whole Numbers

Directions: Fill in the circle beside the best answer for each question.

1. What is the value of the tens and ones blocks?

- ○ **A.** 64
- ○ **B.** 21
- ○ **C.** 46

2. Which number is equal to eight hundred sixty?

- ○ **D.** 816
- ○ **E.** 860
- ○ **F.** 80,060

3. Which number is equal to five thousand seventeen?

- ○ **A.** 500,017
- ○ **B.** 517
- ○ **C.** 5017

4. What is the value of the 3 in 43,682?

- ○ **D.** 30
- ○ **E.** 300
- ○ **F.** 3000

5. Which number is between 461 and 476?

- ○ **A.** 468
- ○ **B.** 495
- ○ **C.** 489

6. An Internet board-game site shows how many games are being played. At 8:30 P.M., which game had the most people playing?

- ○ **D.** Backgammon 665
- ○ **E.** Chess 662
- ○ **F.** Checkers 658

7. 2846 > _____

- ○ **A.** 2851
- ○ **B.** 2902
- ○ **C.** 2837

8. What position in the alphabet is the letter F?

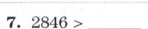

A B C D E F G H I J K . . .

- ○ **D.** fourth
- ○ **E.** sixth
- ○ **F.** eighth

PRACTICE ANSWERS
Page 10

1. 4 tens
2. 8 ones
3. sixty-two
4. nine hundred seventy-five
5.

Hundreds	Tens	Ones
8	4	3

6.

Thousands	Hundreds	Tens	Ones
7	2	5	6

7.

Thousands	Hundreds	Tens	Ones
3	0	1	8

8. 257
9. 1,138 or 1138
10. 3,050 or 3050

Page 11

11. six hundred thousand or 600,000
12. two million or 2,000,000
13. ninety million or 90,000,000
14. <
15. >
16. <
17. 28, 29, 73
18. 2561, 2378, 957
19. first
20. fourth
21. twenty-fifth
22. thirty-second
23. Adam
24. sixth

TEST PREP ANSWERS
Page 12

1. C
2. E
3. C
4. F
5. A
6. D
7. C
8. E

Name _____ Date _____

Money

17–19 Write the amount with a dollar sign and a decimal point.

1.

_____

2.

_____

3.

_____

4.

_____

5.

_____

6. two $20 bills, one $10 bill, three $5 bills, three quarters, one nickel, one penny _____

7. three $10 bills, two $5 bills, three $1 bills, five dimes, six nickels _____

8. Show 2 different ways to make 67¢. Write the number and name of each coin.

_____

_____

20 Write what you would say for each bill or coin when counting the money.

9. Show how to count out $7.53.

$5 $6   _____ _____   _____   _____ $7.53

Name _____ Date _____

The most common way for a clerk to count change is to count up.

10. You owe $4.79. You give the clerk a $5 bill. Count up to
show how the clerk should give you your change. Begin by
saying the amount owed, $4.79. What should the clerk say
while giving you coins to get to $5.00?

$4.79 →

_____   _____   _____

11. You owe $6.63. You give the clerk a $10 bill. Count up
to show how the clerk should give you your change.

   a. Begin by saying the amount owed, $6.63. What should
the clerk say while giving you coins to the next dollar?

$6.63 →

_____   _____   _____   _____

   b. What should the clerk say while giving you bills to reach the
$10 you gave him?

$7.00 → [bills]

_____   _____   _____

12. Draw a picture of the bills and coins you should receive
as change if you give the clerk a $10 bill for a $7.73 purchase.
Draw the smallest value coins first. Draw the bills last.
Write what the clerk would say as she gives you each coin
or bill.

$7.73 →_____   _____   _____   _____   _____

Name _____   Date _____

Money

Directions: Mark the letter beside the best answer to each question.

1. What is the value of the coins?

- ○ **A.** $0.92¢
- ○ **B.** 8¢
- ○ **C.** 92¢
- ○ **D.** 0.92¢

2. What is the value of the coins?

- ○ **F.** $0.76¢
- ○ **G.** $0.76
- ○ **H.** $1.01
- ○ **J.** $76¢

3. What is the value of the money?
two $5 bills, three $1 bills, two quarters, one dime

- ○ **A.** $13.60
- ○ **B.** $10.40
- ○ **C.** $13.55
- ○ **D.** $25.55

4. What is the value of the money?
one $20 bill, three $10 bills, two $5 bills, one quarter, three nickels, four pennies

- ○ **F.** $55.44
- ○ **G.** $60.54
- ○ **H.** $35.29
- ○ **J.** $60.44

5. Jenny owes 68¢. She gives the clerk a $1 bill. What does the clerk give her for change?

- ○ **A.** 2 pennies, a nickel, and a quarter
- ○ **B.** 2 pennies and 2 dimes
- ○ **C.** 2 nickels and a quarter
- ○ **D.** a penny, a nickel, a dime, and a quarter

6. Sean owes $3.98. He gives the clerk a $5 bill. What does the clerk give him for change?

- ○ **F.** two pennies and two $1 bills
- ○ **G.** two pennies, a dime, and a $1 bill
- ○ **H.** two pennies and $1 bill
- ○ **J.** 3 pennies and $1 bill

PRACTICE ANSWERS
Page 14

1. $0.30
2. $1.50
3. $0.68
4. $2.65
5. $7.91
6. $65.81
7. $43.80
8. Some possible answers:
   2 quarters, 1 dime,
   1 nickel, 2 pennies

   2 quarters, 3 nickels,
   2 pennies

   6 dimes, 7 pennies

   4 dimes, 5 nickels,
   2 pennies

   67 pennies
9. $7, $7.25, $7.50, $7.51,
   $7.52

Page 15

10. $4.80, $4.90, $5.00
11. a. $6.64, $6.65, $6.75,
       $7.00

    b. $8.00, $9.00, $10.00
12. Check students'
    drawings: 2 pennies, 1
    quarter, 2 $1-dollar bills
    $7.74, $7.75, $8.00,
    $9.00, $10.00

TEST PREP ANSWERS
Page 16

1. C
2. G
3. A
4. J
5. A
6. H

Name _____ Date _____

Decimals

22–25   Write the value of the shaded part as a decimal.

1. _____

2. _____   **3.** _____

4. _____

Underline the tenths digit.

5. 3.6   **6.** 7.35   **7.** 80.4   **8.** 0.52   **9.** 205.49

What is the value of the 8 in each number?

10. 8.3 _____   **11.** 4.82 _____   **12.** 5.28 _____

13. What is the value of each digit in 3.62?

3 _____   6 _____   2 _____

26   **Write:** 0.4 = 0.40
Say: *Four tenths is equal to forty hundredths.*

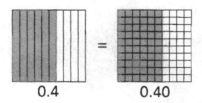

0.4   0.40

Write a decimal equivalent to each number.

14. 0.50 _____   **15.** 0.7 _____   **16.** 4.80 _____   **17.** 3.6 _____

Name _____ Date _____

18.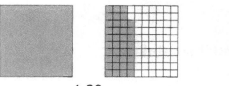

1.38   1.6

Which is greater, 1.38 or 1.6? _____

Write >, <, or = in each ◯.

19. 0.4 ◯ 0.6   **20.** 0.74 ◯ 0.59   **21.** 3.47 ◯ 4.02

22. 0.8 ◯ 0.87   **23.** 0.3 ◯ 0.30   **24.** 21.67 ◯ 21.64

25. Write 0.24, 0.12, 0.15 in order from greatest to least.

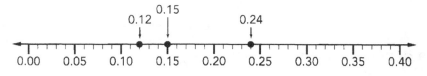

_____   _____   _____

26. Write in order from least to greatest: 0.46 m, 0.51 m, 0.42 m.

_____   _____   _____

Write the decimal as a fraction.

0.20 = $\frac{20}{100}$ = $\frac{2}{10}$ = $\frac{1}{5}$

27. 0.7 _____   **28.** 0.9 _____   **29.** 0.37 _____   **30.** 0.81 _____

Write the decimal as a percent.

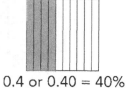

0.4 or 0.40 = 40%   0.83 = 83%

31. 0.70 _____   **32.** 0.25 _____   **33.** 0.8 _____   **34.** 0.5 _____

Name _____ Date _____

Decimals

Write the decimal and fraction to name each shaded part.

1.

_____ _____

2.

_____ _____

3.

_____ _____

4.

_____ _____

5. Write a decimal equivalent to 0.7. _____

Write >, <, or = in each ◯.

6. 0.8 ◯ 0.3   **7.** 0.38 ◯ 0.54   **8.** 6.25 ◯ 6.09

9. 0.7 ◯ 0.70   **10.** 0.6 ◯ 0.62   **11.** 1.21 ◯ 1.08

12. Write 0.42, 0.39, 0.37 in order from least to greatest.

_____ _____ _____

Write the decimal and percent to name each shaded part.

14.

_____ _____

15.

_____ _____

16.

_____ _____

PRACTICE ANSWERS
Page 18
1. 1.35
2. 0.6
3. 0.24
4. 2.07
5. 3.6
6. 7.35
7. 80.4
8. 0.52
9. 205.49
10. 8 or 8 ones
11. 0.8 or 8 tenths
12. 0.08 or 8 hundredths
13. 3 or 3 ones, 6 tenths, 2 hundredths
14. 0.5
15. 0.70
16. 4.8
17. 3.60
18. 1.6
19. <
20. >
21. <
22. <
23. =
24. >

Page 19
25. 0.24, 0.15, 0.12
26. 0.42 m, 0.46 m, 0.51 m
27. $\frac{7}{10}$
28. $\frac{9}{10}$
29. $\frac{37}{100}$
30. $\frac{81}{100}$
31. 70%
32. 25%
33. 80%
34. 50%

TEST PREP ANSWERS
Page 20
1. 0.3, $\frac{3}{10}$
2. 0.9, $\frac{9}{10}$
3. 0.57, $\frac{57}{100}$
4. 0.41, $\frac{41}{100}$
5. 0.70
6. >
7. <
8. >
9. =
10. <
11. >
12. 0.37, 0.39, 0.42
14. 0.50 or 0.5, 50%
15. 0.03, 3%
16. 0.75, 75%

Places for Pets

OBJECTIVES
- Use place value vocabulary and concepts to identify and compare whole numbers
- Use clues to solve number puzzles

MATERIALS
none

TIME
- 20–30 minutes

TEACHER NOTES
- This activity gives students practice using place value vocabulary and concepts. The clues will help students find the answers. They may need help with some terms, such as "digit." Suggest that they draw a line segment for each digit. For example:

  1. _____ _____

  2. _____ _____ _____ _____

- If students use the information for each clue as it is given, they can avoid getting confused with too much information.

ANSWERS
1. 72
2. 4000
3. 10
4. 62,400,000

EXTENSIONS
- Have students make up their own number puzzles using real data on any topic. They can then have other students solve the puzzles.

- Play this game with the class. You will need a set of digit cards labeled 0-9. Write the following on the board and have the students copy it.

  _____ _____ , _____ _____ _____

  Tell the students that you are going to randomly select one digit card for each space. They must write the digit in one of their spaces before you select the next number. "No saving numbers or changing places!" Continue until the number of digits you have selected matches the number of spaces they have copied. The winner is the student(s) with the greatest number. You can repeat the game using more places or including decimal places also.

Name _____ Date _____

Places for Pets

1. Use the clues to find out the number of spots on Pongo in the original *101 Dalmations* movie.
- The number is a 2-digit number.
- The ones digit is between 1 and 3.
- The tens digit is between 6 and 8.

Pongo had _____ spots.

2. Use the clues to find out about how many years ago cats were first domesticated in Ancient Egypt.
- The number is a 4-digit number.
- The ones, tens, and hundreds digits are each zero.
- The number is more than 3000 and less than 5000.

Cats were domesticated in Ancient Egypt about

_____ years ago.

3. Cats have over 100 vocal sounds. Use the clues to find out how many vocal sounds dogs have.
- The number is a 2-digit number.
- Both digits are less than 2.
- The tens digit is one more than the ones digit.

Dogs have _____ vocal sounds.

4. Use the clues to find out about how many dogs are owned in the United States.
- The ones, tens, hundreds, thousands, and ten-thousands digits are each zero.
- A 2 is in the millions place.
- The digit in the hundred-thousands place is two more than the millions digit.
- The digit in the ten-millions place is four more than the millions digit.

There are approximately_____ owned dogs in the United States.

Place Value:
Whole Numbers
pages 2-11

Comparing Whole Numbers
pages 12-13

Name _____ Date _____

Addition Concepts and Facts

36–39

To find a sum, you can
- Draw pictures to show each addend.
- Use counters to model each addend.
- Use a number line. (Find the greater addend.
  The other addend tells you how many jumps to take to the right.)
- Count on from the greater number.

Find each sum.

1. $3 + 5 = $ _____ **2.** $9 + 6 = $ _____ **3.** $4 + 8 = $ _____ **4.** $7 + 5 = $ _____

5. 6   **6.** 4   **7.** 7   **8.** 8
     +4   +9   +2   +4

Use the addition table to find each sum.

+	4	5	6	7	8	9
4	8	9	10	11	12	13
5	9	10	11	12	13	14
6	10	11	12	13	14	15
7	11	12	13	14	15	16

9. $8 + 5 = $ _____

10. $9 + 4 = $ _____

11. $7 + 7 = $ _____

12. $7 + 5 = $ _____

40

When you add zero to a number, the sum is that number.

Add.

13. $0 + 9 = $ ___ **14.** $7 + 0 = $ ___ **15.** $4 + 0 = $ ___ **16.** $0 + 8 = $ ___

41

If you change the order of two addends, the sum stays the same.

Add.

17. $6 + 3 = $ _____   **18.** $4 + 7 = $ _____   **19.** $8 + 5 = $ _____

$3 + 6 = $ _____ $7 + 4 = $ _____ $5 + 8 = $ _____

Name _____ Date _____

When you add 1 or 2 to a number, it's like counting.

Add.

20.   6   **21.**   6   **22.**   1   **23.**   2
        +1   +2   +8   +8

A doubles addition fact has two addends that are the same.   42–43
Sometimes you can use doubles facts to find other facts.

Find each sum.

24. 4 + 4 = _____   **25.** 7 + 7 = _____   **26.** 8 + 8 = _____

   4 + 5 = _____   7 + 8 = _____   9 + 8 = _____

27. Write 6 addition sentences that have a sum of 10.   44

   _____ + _____ = 10   _____ + _____ = 10   _____ + _____ = 10

   _____ + _____ = 10   _____ + _____ = 10   _____ + _____ = 10

You can use facts with sums of 10 to find sums greater than 10.   45

7 + 4 = ▧
Think of 4 as 3 + 1.   7 + 3 = 10 and 1 more is 11.

7 + 4 = 11

28. 7 + 6 = _____   **29.** 7 + 5 = _____   **30.** 7 + 8 = _____

9 + 3 = ▧
Think of 3 as 1 + 2.   9 + 1 = 10 and 2 more is 12.

9 + 3 = 12

31. 9 + 2 = _____   **32.** 9 + 5 = _____   **33.** 9 + 7 = _____

Name _____   Date _____

Addition Concepts and Facts

Circle the letter of the correct answer.

1. 4 + 3

   A. 5

   B. 7

   C. 9

   D. 8

2. 7 + 8

   A. 13

   B. 16

   C. 14

   D. 15

3. 9 + 6

   A. 14

   B. 15

   C. 16

   D. 17

4. 8 + 9

   A. 10

   B. 15

   C. 16

   D. 17

5. Write four addition sentences that have a sum of 9.

   ____ + ____ = 9

   ____ + ____ = 9

   ____ + ____ = 9

   ____ + ____ = 9

6. What is the sum when you add zero to a number?

   _____

   _____

7. Write a number sentence to show how many stamps the girl will have in all.

   _____

PRACTICE ANSWERS
Page 24
1. 8
2. 15
3. 12
4. 12
5. 10
6. 13
7. 9
8. 12
9. 13
10. 13
11. 14
12. 12
13. 9
14. 7
15. 4
16. 8
17. 9, 9
18. 11, 11
19. 13, 13

Page 25
20. 7
21. 8
22. 9
23. 10
24. 8, 9
25. 14, 15
26. 16, 17
27. Any four of the following addition sentences:
$0 + 10 = 10$
$1 + 9 = 10$
$2 + 8 = 10$
$3 + 7 = 10$
$4 + 6 = 10$
$5 + 5 = 10$
$6 + 4 = 10$
$7 + 3 = 10$
$8 + 2 = 10$
$9 + 1 = 10$
$10 + 0 = 10$
28. 13
29. 12
30. 15
31. 11
32. 14
33. 16

TEST PREP ANSWERS
Page 26
1. B
2. D
3. B
4. D
5. Any four of the following addition sentences:
$0 + 9 = 9$
$1 + 8 = 9$
$2 + 7 = 9$
$3 + 6 = 9$
$4 + 5 = 9$
$5 + 4 = 9$
$6 + 3 = 9$
$7 + 2 = 9$
$8 + 1 = 9$
$9 + 0 = 9$
6. When you add zero to a number, the sum is that same number.
7. $8 + 3 = 11$

Name _____ Date _____

Subtraction Concepts and Facts

46–49 **When you see subtraction, you can think about addition.**

1. Complete.

Addition Facts	Related Subtraction Facts
$3 + 8 = 11$	$11 - 8 =$ _____
$8 + 3 = 11$	$11 - 3 =$ _____

Find the missing addend.

2. $6 +$ _____ $= 14$ 　　　　　**3.** _____ $+ 9 = 17$

4. $8 +$ _____ $= 12$ 　　　　　**5.** _____ $+ 5 = 13$

50–52 **To subtract, you can**

• Use counters to model the action or comparison.
• Use a number line. (Start at the first number. The number you are subtracting tells you how many jumps to take to the left.)
• Count back from the greater number.

Find the difference.

6. $7 - 2 =$ ___ 　　**7.** $9 - 5 =$ ___ 　　**8.** $12 - 8 =$ ___ 　　**9.** $15 - 6 =$ ___

10. $\begin{array}{r} 9 \\ -3 \\ \hline \end{array}$ 　　　**11.** $\begin{array}{r} 10 \\ -6 \\ \hline \end{array}$ 　　　**12.** $\begin{array}{r} 12 \\ -8 \\ \hline \end{array}$ 　　　**13.** $\begin{array}{r} 15 \\ -6 \\ \hline \end{array}$

53 Use the addition table to find the difference.

14. $9 - 5 =$ _____

15. $11 - 4 =$ _____

16. $15 - 7 =$ _____

17. $16 - 9 =$ _____

+	4	5	6	7	8	9
4	8	9	10	11	12	13
5	9	10	11	12	13	14
6	10	11	12	13	14	15
7	11	12	13	14	15	16

Name _____ Date _____

Most fact families consist of four related facts. Each fact uses the same three numbers.

54–55

18. Write two addition facts and two subtraction facts using 13, 5, and 8.

Addition Facts   Subtraction Facts

_____   _____

_____   _____

When you subtract zero from a number, the difference is the number you subtracted from.

56

Subtract.

19. $5 - 0 =$ ___ **20.** $7 - 0 =$ ___ **21.** $8 - 0 =$ ___ **22.** $0 - 0 =$ ___

When you subtract a number from itself, the difference is 0.

Subtract.

23. $4 - 4 =$ ___ **24.** $9 - 9 =$ ___ **25.** $1 - 1 =$ ___ **26.** $6 - 6 =$ ___

When you subtract 1 or 2 from a number, it's like counting back 1 or 2.

Subtract.

27. $4 - 1 =$ ___ **28.** $9 - 2 =$ ___ **29.** $7 - 1 =$ ___ **30.** $6 - 2 =$ ___

If you know addition doubles, then you also know the related subtraction doubles.

57

31. Complete.

Addition Doubles   Subtraction Doubles

$7 + 7 =$ _____ $14 - 7 =$ _____

$9 + 9 =$ _____ $18 - 9 =$ _____

Find the difference.

58–59

32. $10 - 4 =$ ___ **33.** $10 - 7 =$ ___ **34.** $10 - 5 =$ ___ **35.** $14 - 9 =$ ___

36. $12 - 9 =$ ___ **37.** $12 - 4 =$ ___ **38.** $16 - 9 =$ ___ **39.** $16 - 8 =$ ___

Name _____   Date _____

Subtraction Concepts and Facts

Directions: Fill in the circle next to the letter of the correct answer.

1. 8 − 3

○ **A.** 11

○ **B.** 4

○ **C.** 6

○ **D.** 5

2. 10 − 6

○ **A.** 6

○ **B.** 4

○ **C.** 5

○ **D.** 3

3. 12 − 9

○ **E.** 6

○ **F.** 4

○ **G.** 5

○ **H.** 3

4. 16 − 8

○ **A.** 6

○ **B.** 9

○ **C.** 8

○ **D.** 7

5. You have 7 marbles. You lose 3 marbles. How many do you have now?

○ **A.** 10

○ **B.** 4

○ **C.** 3

○ **D.** 11

6. You have 7 marbles. You win 3 more marbles. How many do you have now?

○ **E.** 10

○ **F.** 4

○ **G.** 3

○ **H.** 11

7. You have 7 marbles. Al has 3 marbles. How many more marbles do you have than Al?

○ **A.** 10

○ **B.** 4

○ **C.** 3

○ **D.** 11

PRACTICE ANSWERS
Page 28
1. 3, 8
2. 8
3. 8
4. 4
5. 8
6. 5
7. 4
8. 4
9. 9
10. 6
11. 4
12. 4
13. 9
14. 4
15. 7
16. 8
17. 7

Page 29
18. $5 + 8 = 13$ $13 - 5 = 8$
    $8 + 5 = 13$ $13 - 8 = 5$
19. 5
20. 7
21. 8
22. 0
23. 0
24. 0
25. 0
26. 0
27. 3
28. 7
29. 6
30. 4
31. $7 + 7 = \underline{14}$ $14 - 7 = \underline{7}$
    $9 + 9 = \underline{18}$ $18 - 9 = \underline{9}$
32. 6
33. 3
34. 5
35. 5
36. 3
37. 8
38. 7
39. 8

TEST PREP ANSWERS
Page 30
1. D
2. B
3. H
4. C
5. B
6. E
7. B

Name _____ Date _____

Multiplication Concepts and Facts

60–61 Multiplication is a shortcut for addition. When you have equal amounts to add, you can multiply to find the total.

You can multiply to join equal groups to find a total.

1. How many groups of stars are there? _____

2. How many stars are there in each group? _____

3. How many stars in all? _____

4. $2 \times 3 =$ _____

5. What two numbers would you multiply to find out how many tires are on 3 bicycles? _____ _____

You can multiply equal money amounts to find a total.

6. What two numbers would you multiply to find how much it will cost to buy 3 candles that cost 5¢ each? _____ _____

You can multiply to find the number of objects in an array.

7. How many are in each row? _____

8. How many are in each column? _____

9. How many in all? _____

10. $2 \times 4 =$ _____

11. A parking lot has 3 rows of cars. There are 4 cars in each row. How many cars are in the parking lot? _____

Name _____ Date _____

Use the multiplication table to find the product.

×	4	5	6	7	8	9
4	16	20	24	28	32	36
5	20	25	30	35	40	45
6	24	30	36	42	48	54
7	28	35	42	49	56	63

12. $4 \times 5 =$ _____

13. $6 \times 6 =$ _____

14. $8 \times 7 =$ _____   **15.** $7 \times 9 =$ _____   **16.** $9 \times 6 =$ _____

17. $9 \times 8 =$ _____   **18.** $7 \times 5 =$ _____   **19.** $4 \times 8 =$ _____

20. $5 \times 5 =$ _____   **21.** $6 \times 4 =$ _____   **22.** $6 \times 8 =$ _____

To multiply, you can
- Draw a picture.
- Use a number line. (Skip count.)
- Use repeated addition.
- Make or draw an array.

Find the product.

23. $4 \times 2 =$ _____   **24.** $3 \times 8 =$ _____   **25.** $4 \times 4 =$ _____

26. $6 \times 4 =$ _____   **27.** $5 \times 7 =$ _____   **28.** $7 \times 3 =$ _____

When one factor is 1, the product is the same as the other factor.

Find the product.

29. $1 \times 7 =$ _____   **30.** $9 \times 1 =$ _____   **31.** $1 \times 6 =$ _____

32. $1 \times 4 =$ _____   **33.** $5 \times 1 =$ _____   **34.** $2 \times 1 =$ _____

35. $7 \times 1 =$ _____   **36.** $8 \times 1 =$ _____   **37.** $1 \times 1 =$ _____

38. $1 \times 3 =$ _____   **39.** $1 \times 8 =$ _____   **40.** $1 \times 9 =$ _____

Name _____ Date _____

67 **When one factor is 0, the product is 0.**

Find the product.

41. $0 \times 3 =$ ___ **42.** $0 \times 8 =$ ___ **43.** $1 \times 0 =$ ___ **44.** $0 \times 6 =$ ___

45. $5 \times 0 =$ ___ **46.** $0 \times 3 =$ ___ **47.** $0 \times 9 =$ ___ **48.** $7 \times 0 =$ ___

If you change the order of two factors, or *turn them around*, the product is the same.

Find the product.

49. $4 \times 2 =$ ____ **50.** $3 \times 9 =$ ____ **51.** $5 \times 8 =$ ____

$2 \times 4 =$ ____ $9 \times 3 =$ ____ $8 \times 5 =$ ____

68 **You can use what you know about adding doubles to multiply by 2.**

Find the product.

52. $2 \times 6 =$ ____ **53.** $8 \times 2 =$ ____ **54.** $2 \times 7 =$ ____

55. $5 \times 2 =$ ____ **56.** $0 \times 2 =$ ____ **57.** $1 \times 2 =$ ____

Knowing doubles can help you multiply by 3.
 To multiply 4 by 3:
 Double 4. **Then add 4 more.**
 $4 + 4 = 8$ **$8 + 4 = 12$**

 So, $3 \times 4 = 12$

Try the double and add more method to find the product..

58. $3 \times 5 =$ ____ **59.** $3 \times 3 =$ ____ **60.** $3 \times 6 =$ ____

61. $3 \times 1 =$ ____ **62.** $0 \times 3 =$ ____ **63.** $9 \times 3 =$ ____

Name _____ Date _____

Knowing doubles can help you multiply by 4.
    To multiply 6 by 4:

Double 6.	**Double again.**
$6 + 6 = 12$	$12 + 12 = 24$

    So, $6 \times 4 = 24$

Try the double and then double again method to find the product.

64. $4 \times 5 =$ ____   **65.** $4 \times 3 =$ ____   **66.** $7 \times 4 =$ ____

67. $4 \times 4 =$ ____   **68.** $2 \times 4 =$ ____   **69.** $4 \times 5 =$ ____

70. $1 \times 4 =$ ____   **71.** $9 \times 4 =$ ____   **72.** $6 \times 4 =$ ____

73. Skip count by 5 to 40. ____ ____ ____ ____ ____ ____ ____ ____

    What patterns do you see? _____

_____

These patterns can make the fives facts easy.

You can also think about nickels when you multiply by 5.

74. 2 nickels = _____   **75.** 3 nickels = _____   **76.** 6 nickels = _____

    $2 \times 5 =$ _____ $3 \times 5 =$ _____ $6 \times 5 =$ _____

77. $7 \times 5 =$ _____   **78.** $5 \times 4 =$ _____   **79.** $0 \times 5 =$ _____

80. $5 \times 8 =$ _____   **81.** $5 \times 5 =$ _____   **82.** $5 \times 1 =$ _____

83. $5 \times 2 =$ _____   **84.** $5 \times 9 =$ _____   **85.** $8 \times 5 =$ _____

86. $5 \times 7 =$ _____   **87.** $5 \times 6 =$ _____   **88.** $5 \times 3 =$ _____

Name _____ Date _____

70-71

89. Skip count by 9 to 81. ____ ____ ____ ____ ____ ____ ____ ____ ____

What patterns do you see? _____

_____

_____

Use the patterns to find these products

90. $9 \times 3 =$ _____   **91.** $9 \times 6 =$ _____   **92.** $0 \times 9 =$ _____

93. $9 \times 7 =$ _____   **94.** $2 \times 9 =$ _____   **95.** $9 \times 9 =$ _____

96. $9 \times 4 =$ _____   **97.** $9 \times 1 =$ _____   **98.** $9 \times 8 =$ _____

99. $9 \times 6 =$ _____   **100.** $5 \times 9 =$ _____   **101.** $3 \times 9 =$ _____

72-73 **You can use facts you know to find unknown facts.**

Doubling

102. $6 \times 4 = 24$, so $6 \times 8 =$ _____

103. $7 \times 3 = 21$, so $7 \times 6 =$ _____

104. $9 \times 2 = 18$, so $9 \times 4 =$ _____

105. $8 \times 3 = 24$, so $8 \times 6 =$ _____

106. $3 \times 4 = 12$, so $3 \times 8 =$ _____

Adding one more set

107. $5 \times 8 = 40$, so $6 \times 8 =$ _____

108. $6 \times 6 = 36$, so $7 \times 6 =$ _____

109. $8 \times 4 = 32$, so $9 \times 4 =$ _____

110. $7 \times 6 = 42$, so $8 \times 6 =$ _____

111. $3 \times 7 = 21$, so $3 \times 8 =$ _____

Name _____ Date _____

Multiplication Concepts and Facts

Directions: Fill in the circle beside the letter of the correct answer.

1. Which picture can help you find the product of 3 and 4?

   ○ **A.** ☆☆☆
            ☆☆☆

   ○ **B.** ☆☆☆☆
            ☆☆☆☆

   ○ **C.** ☆☆☆☆
            ☆☆☆☆
            ☆☆☆☆

2. Which has the same value as 4×6?

   ○ **A.** $4 + 6$

   ○ **B.** $6 + 6 + 6 + 6$

   ○ **C.** $4 + 4 + 4 + 4 + 4$

3. Which set of factors has a product of 15?

   ○ **A.** 5 and 3

   ○ **B.** 5 and 10

   ○ **C.** 7 and 8

4. Which is another way to write 4 groups of 7?

   ○ **A.** 4×7

   ○ **B.** $4 + 7$

   ○ **C.** $7 - 4$

5. What is another way to find 6×8?

   ○ **A.** $6 + 8$

   ○ **B.** 6×0 and add 8 more

   ○ **C.** 8×6

6. Tickets cost 5¢ each. How much will 6 tickets cost?

   ○ **A.** 11¢

   ○ **B.** 1¢

   ○ **C.** 30¢

7. There are 4 chairs at each table. There are 9 tables. How many chairs are there in all?

   ○ **A.** 49 chairs

   ○ **B.** 36 chairs

   ○ **C.** 13 chairs

8. 7 plants are in each row. There are 6 rows. How many plants are there?

   ○ **A.** 13 plants

   ○ **B.** 76 plants

   ○ **C.** 42 plants

PRACTICE ANSWERS
Page 32
1. 2
2. 3
3. 6
4. 6
5. 2 and 3
6. 3 and 5
7. 4
8. 2
9. 8
10. 8
11. 12

Page 33
12. 20
13. 36
14. 56
15. 63
16. 54
17. 72
18. 35
19. 32
20. 25
21. 24
22. 48

23. 8	24. 24	25. 16
26. 24	27. 35	28. 21
29. 7	30. 9	31. 6
32. 4	33. 5	34. 2
35. 7	36. 8	37. 1
38. 3	39. 8	40. 9

Page 34
41. 0
42. 0
43. 0
44. 0
45. 0
46. 0

47. 0
48. 0
49. 8, 8
50. 27, 27
51. 40, 40
52. 12
53. 16
54. 14
55. 10
56. 0
57. 2
58. 15
59. 9
60. 18
61. 3
62. 0
63. 27

Page 35
64. 20
65. 12
66. 28
67. 16
68. 8
69. 20
70. 4
71. 36
72. 24
73. 5, 10, 15, 20, 25, 30, 35, 40
   The ones digits have a pattern : 0, 1, 0, 1, …
   The tens digits have a pattern: 0, 1, 1, 2, 2, 3, 3, …
74. 10¢, 10
75. 15¢, 15
76. 30¢, 30

77. 35	78. 20	79. 0
80. 40	81. 25	82. 5
83. 10	84. 45	85. 40
86. 35	87. 30	88. 15

Page 36
89. 9, 18, 27, 36, 45, 54, 63, 72, 81
   The tens digit in the product is always one less than the number you are multiplying by 9. The sum of the two digits in the product is always 9.

90. 27	91. 54	92. 0
93. 63	94. 18	95. 81
96. 36	97. 9	98. 72
99. 54	100. 45	101. 27

102. 48
103. 42
104. 36
105. 48
106. 24
107. 48
108. 42
109. 36
110. 48
111. 24

TEST PREP ANSWERS
Page 37
1. C
2. B
3. A
4. A
5. C
6. C
7. B
8. C

Name _____ Date _____

Division Concepts and Facts

You can use what you know about multiplication to help you with division.

You can divide to find the number in each group.

1. How many photos in all? _____

2. How many groups? _____

3. How many in each group? _____

You can divide to find the number of groups.

4. How many muffins in all? _____

5. How many muffins are on each plate? _____

6. How many plates are there? _____

**Sometimes you can't divide an amount into equal groups.
The left over amount is the remainder.**

You have 7 cookies to share among 3 people.

7. How many does each person get? _____

8. How many are left over? _____

There are two common ways to write division.   76

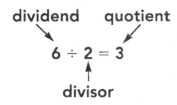

dividend   quotient

6 ÷ 2 = 3

divisor

9. Label this form of division.

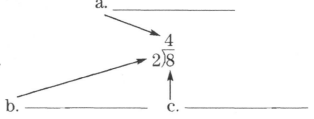

a. _____

$2\overline{)8}$ with 4

b. _____   c. _____

Name _____ Date _____

77–80 **Division and multiplication are related.**

10. Complete.

Multiplication Facts   Related Division Facts

$3 \times 4 = 12$ $12 \div 4 =$ _____

$6 \times 3 = 18$ $18 \div 3 =$ _____

You can use multiplication facts to help you with division.
Find the missing factors to help you find the quotients.

11. _____ $\times 4 = 12$ **12.** _____ $\times 6 = 42$ **13.** _____ $\times 9 = 18$

     $12 \div 4 =$ _____ $42 \div 6 =$ _____ $18 \div 9 =$ _____

14. $8 \times$ _____ $= 24$ **15.** $3 \times$ _____ $= 15$ **16.** $7 \times$ _____ $= 21$

     $24 \div 8 =$ _____ $15 \div 3 =$ _____ $21 \div 7 =$ _____

81 Use the multiplication table to divide.

$\times$	4	5	6	7	8	9
4	16	20	24	28	32	36
5	20	25	30	35	40	45
6	24	30	36	42	48	54
7	28	35	42	49	56	63

17. $20 \div 4 =$ _____   **18.** $24 \div 6 =$ _____   **19.** $42 \div 7 =$ _____

20. $30 \div 5 =$ _____   **21.** $35 \div 7 =$ _____   **22.** $36 \div 6 =$ _____

23. $54 \div 6 =$ _____   **24.** $56 \div 8 =$ _____   **25.** $32 \div 4 =$ _____

26. $49 \div 7 =$ _____   **27.** $63 \div 9 =$ _____   **28.** $28 \div 4 =$ _____

Name _____ Date _____

You can use repeated subtraction to find a quotient.

82

29. $20 \div 5 =$ _____

$$\begin{array}{r} 20 \\ -\ 5 \\ \hline 15 \\ -\ 5 \\ \hline 10 \\ -\ 5 \\ \hline 5 \\ -\ 5 \\ \hline 0 \end{array}$$

Count how many times you subtracted 5.

Fact families can help you learn division facts.

82–83

Most families have four related facts. Each fact uses the same three numbers.

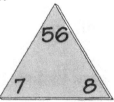

Complete each fact family.

30. $7 \times 8 = 56$ $56 \div 8 =$ _____

     $8 \times 7 =$ _____ $56 \div 7 =$ _____

31. $6 \times$ _____ $= 48$ $48 \div 8 =$ _____

     $8 \times 6 =$ _____ $48 \div 6 =$ _____

32. Write the family of facts for the numbers 4, 7, and 28.

_____   _____

_____   _____

Name _____ Date _____

84 **When you divide any number by 1, you get that same number.**

Find each quotient.

33. $6 \div 1 =$ ___ **34.** $4 \div 1 =$ ___ **35.** $7 \div 1 =$ ___ **36.** $9 \div 1 =$ ___

When you divide a number other than zero by itself, you get 1.

Divide.

37. $5 \div 5 =$ ___ **38.** $8 \div 8 =$ ___ **39.** $3 \div 3 =$ ___ **40.** $6 \div 6 =$ ___

When you divide zero by any number, you get 0.
You cannot divide a number by zero.

Write the quotient, if you can divide. If you cannot divide, cross the expression out.

41. $0 \div 5 =$ ___ **42.** $0 \div 8 =$ ___ **43.** $6 \div 0 =$ ___ **44.** $0 \div 2 =$ ___

85 **When you divide a number by 2, you are finding half of the number.**

Find each quotient.

45. $8 \div 2 =$ ___ **46.** $16 \div 2 =$ ___ **47.** $6 \div 2 =$ ___

48. $4 \div 2 =$ ___ **49.** $10 \div 2 =$ ___ **50.** $18 \div 2 =$ ___

When you divide a number by 4, you can divide it by 2 and then divide that quotient by 2.

Find each quotient.

51. $20 \div 4 =$ ___ **52.** $8 \div 4 =$ ___ **53.** $12 \div 4 =$ ___

54. $16 \div 4 =$ ___ **55.** $20 \div 4 =$ ___ **56.** $32 \div 4 =$ ___

Name _____ Date _____

Division Concepts and Facts

Directions: Write each answer.

Write a number in each blank.

1. $7 \times \underline{\hspace{1cm}} = 14$ **2.** $3 \times \underline{\hspace{1cm}} = 12$ **3.** $5 \times \underline{\hspace{1cm}} = 45$

$14 \div 7 = \underline{\hspace{1cm}}$ $12 \div 3 = \underline{\hspace{1cm}}$ $45 \div 5 = \underline{\hspace{1cm}}$

4. $4 \times \underline{\hspace{1cm}} = 24$ **5.** $6 \times \underline{\hspace{1cm}} = 42$ **6.** $8 \times \underline{\hspace{1cm}} = 48$

$24 \div 4 = \underline{\hspace{1cm}}$ $42 \div 6 = \underline{\hspace{1cm}}$ $48 \div 8 = \underline{\hspace{1cm}}$

Use the words in the Word Bank to fill in the blanks.

Word Bank
divisor
quotient
remainder

7. The answer in division is called the _____.

8. The dividend is divided by the _____.

9. When you can't divide an amount into equal amounts, the leftover amount is called the _____.

Fill in the blanks with the number 1 or 0 .

10. When you divide a number other than zero by itself, you get _____.

11. When you divide 0 by any number you get _____.

12. You cannot divide a number by _____.

13. When you divide any number by _____, you get the number you started with.

Solve each problem.

14. If each car holds 4 children, how many cars will be needed for 12 children? _____

15. 12 children are in the van. 4 children are in the car. How many more children are in the van? _____

PRACTICE ANSWERS
Page 39

1. 6
2. 3
3. 2
4. 8
5. 2
6. 4
7. 2
8. 1
9. a. quotient
   b. divisor
   c. dividend

Page 40

10. 3, 6
11. 3, 3
12. 7, 7
13. 2, 2
14. 3, 3
15. 5, 5
16. 3, 3
17. 5
18. 4
19. 6
20. 6
21. 5
22. 6
23. 9
24. 7
25. 8
26. 7
27. 7
28. 7

Page 41

29. 4
30. $7 \times 8 = 56$ $56 \div 8 = \underline{7}$
    $8 \times 7 = \underline{56}$ $56 \div 7 = \underline{8}$
31. $6 \times \underline{8} = 48$ $48 \div 8 = \underline{6}$
    $8 \times 6 = \underline{48}$ $48 \div 6 = \underline{8}$
32. $4 \times 7 = 28$ $28 \div 7 = 4$
    $7 \times 4 = 28$ $28 \div 4 = 4$

Page 42

33. 6
34. 4
35. 7
36. 9
37. 1
38. 1
39. 1
40. 1
41. 0
42. 0
43. ~~6 × 0~~
44. 0
45. 4
46. 8
47. 3
48. 2
49. 5
50. 9
51. 5
52. 2
53. 3
54. 4
55. 5
56. 8

TEST PREP ANSWERS
Page 43

1. 2, 2
2. 4, 4
3. 9, 9
4. 6, 6
5. 7, 7
6. 6, 6
7. quotient
8. divisor
9. remainder
10. 1
11. 0
12. 0
13. 1
14. 3 cars
15. 8 children

Name _____ Date _____

Mastering Basic Facts

Write the sum. Cross out the facts you know. Keep track
of facts you need to study.

86-87

Add 0	Add 1	Add 2	Add 3	Add 4
0 + 0 = ___	0 + 1 = ___	0 + 2 = ___	0 + 3 = ___	0 + 4 = ___
1 + 0 = ___	1 + 1 = ___	1 + 2 = ___	1 + 3 = ___	1 + 4 = ___
2 + 0 = ___	2 + 1 = ___	2 + 2 = ___	2 + 3 = ___	2 + 4 = ___
3 + 0 = ___	3 + 1 = ___	3 + 2 = ___	3 + 3 = ___	3 + 4 = ___
4 + 0 = ___	4 + 1 = ___	4 + 2 = ___	4 + 3 = ___	4 + 4 = ___
5 + 0 = ___	5 + 1 = ___	5 + 2 = ___	5 + 3 = ___	5 + 4 = ___
6 + 0 = ___	6 + 1 = ___	6 + 2 = ___	6 + 3 = ___	6 + 4 = ___
7 + 0 = ___	7 + 1 = ___	7 + 2 = ___	7 + 3 = ___	7 + 4 = ___
8 + 0 = ___	8 + 1 = ___	8 + 2 = ___	8 + 3 = ___	8 + 4 = ___
9 + 0 = ___	9 + 1 = ___	9 + 2 = ___	9 + 3 = ___	9 + 4 = ___

Add 5	Add 6	Add 7	Add 8	Add 9
0 + 5 = ___	0 + 6 = ___	0 + 7 = ___	0 + 8 = ___	0 + 9 = ___
1 + 5 = ___	1 + 6 = ___	1 + 7 = ___	1 + 8 = ___	1 + 9 = ___
2 + 5 = ___	2 + 6 = ___	2 + 7 = ___	2 + 8 = ___	2 + 9 = ___
3 + 5 = ___	3 + 6 = ___	3 + 7 = ___	3 + 8 = ___	3 + 9 = ___
4 + 5 = ___	4 + 6 = ___	4 + 7 = ___	4 + 8 = ___	4 + 9 = ___
5 + 5 = ___	5 + 6 = ___	5 + 7 = ___	5 + 8 = ___	5 + 9 = ___
6 + 5 = ___	6 + 6 = ___	6 + 7 = ___	6 + 8 = ___	6 + 9 = ___
7 + 5 = ___	7 + 6 = ___	7 + 7 = ___	7 + 8 = ___	7 + 9 = ___
8 + 5 = ___	8 + 6 = ___	8 + 7 = ___	8 + 8 = ___	8 + 9 = ___
9 + 5 = ___	9 + 6 = ___	9 + 7 = ___	9 + 8 = ___	9 + 9 = ___

Name _____ Date _____

86–87 **Find each sum.**

ROW 1

3	5	5	7	4	6	6	4
+9	+0	+4	+7	+8	+1	+7	+3

ROW 2

7	8	7	4	2	8	4	7
+1	+3	+9	+4	+9	+0	+9	+2

ROW 3

8	1	4	5	6	5	4	0
+2	+9	+6	+6	+8	+3	+7	+9

ROW 4

7	5	3	5	5	7	8	9
+5	+9	+6	+2	+8	+6	+6	+6

ROW 5

9	6	3	4	7	8	9	3
+9	+4	+8	+5	+8	+8	+8	+7

ROW 6

9	7	8	9	3	5	6	6
+2	+4	+7	+7	+5	+7	+2	+9

ROW 7

6	7	8	3	4	6	7	9
+6	+3	+5	+3	+2	+5	+0	+4

Name _____ Date _____

Write the difference. Cross out the facts you know.
Keep track of facts you need to study.

Subtract 0	Subtract 1	Subtract 2	Subtract 3	Subtract 4
0 − 0 = ___	1 − 1 = ___	2 − 2 = ___	3 − 3 = ___	4 − 4 = ___
1 − 0 = ___	2 − 1 = ___	3 − 2 = ___	4 − 3 = ___	5 − 4 = ___
2 − 0 = ___	3 − 1 = ___	4 − 2 = ___	5 − 3 = ___	6 − 4 = ___
3 − 0 = ___	4 − 1 = ___	5 − 2 = ___	6 − 3 = ___	7 − 4 = ___
4 − 0 = ___	5 − 1 = ___	6 − 2 = ___	7 − 3 = ___	8 − 4 = ___
5 − 0 = ___	6 − 1 = ___	7 − 2 = ___	8 − 3 = ___	9 − 4 = ___
6 − 0 = ___	7 − 1 = ___	8 − 2 = ___	9 − 3 = ___	10 − 4 = ___
7 − 0 = ___	8 − 1 = ___	9 − 2 = ___	10 − 3 = ___	11 − 4 = ___
8 − 0 = ___	9 − 1 = ___	10 − 2 = ___	11 − 3 = ___	12 − 4 = ___
9 − 0 = ___	10 − 1 = ___	11 − 2 = ___	12 − 3 = ___	13 − 4 = ___

Subtract 5	Subtract 6	Subtract 7	Subtract 8	Subtract 9
5 − 5 = ___	6 − 6 = ___	7 − 7 = ___	8 − 8 = ___	9 − 9 = ___
6 − 5 = ___	7 − 6 = ___	8 − 7 = ___	9 − 8 = ___	10 − 9 = ___
7 − 5 = ___	8 − 6 = ___	9 − 7 = ___	10 − 8 = ___	11 − 9 = ___
8 − 5 = ___	9 − 6 = ___	10 − 7 = ___	11 − 8 = ___	12 − 9 = ___
9 − 5 = ___	10 − 6 = ___	11 − 7 = ___	12 − 8 = ___	13 − 9 = ___
10 − 5 = ___	11 − 6 = ___	12 − 7 = ___	13 − 8 = ___	14 − 9 = ___
11 − 5 = ___	12 − 6 = ___	13 − 7 = ___	14 − 8 = ___	15 − 9 = ___
12 − 5 = ___	13 − 6 = ___	14 − 7 = ___	15 − 8 = ___	16 − 9 = ___
13 − 5 = ___	14 − 6 = ___	15 − 7 = ___	16 − 8 = ___	17 − 9 = ___
14 − 5 = ___	15 − 6 = ___	16 − 7 = ___	17 − 8 = ___	18 − 9 = ___

Name _____ Date _____

86–87   Find each difference.

ROW 1

12	9	17	14	4	8	14	7
−3	−6	−9	−8	−0	−7	−6	−6

ROW 2

13	6	5	9	7	11	3	14
−9	−4	−5	−0	−3	−8	−3	−9

ROW 3

12	13	11	4	9	6	7	13
−6	−8	−5	−4	−8	−3	−7	−5

ROW 4

11	9	6	17	6	11	16	10
−6	−5	−0	−8	−2	−7	−8	−3

ROW 5

15	9	5	14	8	10	8	4
−6	−2	−4	−7	−5	−2	−3	−2

ROW 6

9	15	10	12	8	11	15	10
−9	−8	−8	−7	−4	−3	−7	−9

ROW 7

9	18	1	7	8	13	12	10
−4	−9	−1	−4	−0	−6	−5	−4

ROW 8

10	2	11	13	16	8	9	11
−7	−2	−9	−4	−7	−2	−7	−4

Name _____　Date _____

Write the product. Cross out the facts you know.
Keep track of facts you need to study.

× 0	× 1	× 2	× 3	× 4
0 × 0 = ___	0 × 1 = ___	0 × 2 = ___	0 × 3 = ___	0 × 4 = ___
1 × 0 = ___	1 × 1 = ___	1 × 2 = ___	1 × 3 = ___	1 × 4 = ___
2 × 0 = ___	2 × 1 = ___	2 × 2 = ___	2 × 3 = ___	2 × 4 = ___
3 × 0 = ___	3 × 1 = ___	3 × 2 = ___	3 × 3 = ___	3 × 4 = ___
4 × 0 = ___	4 × 1 = ___	4 × 2 = ___	4 × 3 = ___	4 × 4 = ___
5 × 0 = ___	5 × 1 = ___	5 × 2 = ___	5 × 3 = ___	5 × 4 = ___
6 × 0 = ___	6 × 1 = ___	6 × 2 = ___	6 × 3 = ___	6 × 4 = ___
7 × 0 = ___	7 × 1 = ___	7 × 2 = ___	7 × 3 = ___	7 × 4 = ___
8 × 0 = ___	8 × 1 = ___	8 × 2 = ___	8 × 3 = ___	8 × 4 = ___
9 × 0 = ___	9 × 1 = ___	9 × 2 = ___	9 × 3 = ___	9 × 4 = ___

× 5	× 6	× 7	× 8	× 9
0 × 5 = ___	0 × 6 = ___	0 × 7 = ___	0 × 8 = ___	0 × 9 = ___
1 × 5 = ___	1 × 6 = ___	1 × 7 = ___	1 × 8 = ___	1 × 9 = ___
2 × 5 = ___	2 × 6 = ___	2 × 7 = ___	2 × 8 = ___	2 × 9 = ___
3 × 5 = ___	3 × 6 = ___	3 × 7 = ___	3 × 8 = ___	3 × 9 = ___
4 × 5 = ___	4 × 6 = ___	4 × 7 = ___	4 × 8 = ___	4 × 9 = ___
5 × 5 = ___	5 × 6 = ___	5 × 7 = ___	5 × 8 = ___	5 × 9 = ___
6 × 5 = ___	6 × 6 = ___	6 × 7 = ___	6 × 8 = ___	6 × 9 = ___
7 × 5 = ___	7 × 6 = ___	7 × 7 = ___	7 × 8 = ___	7 × 9 = ___
8 × 5 = ___	8 × 6 = ___	8 × 7 = ___	8 × 8 = ___	8 × 9 = ___
9 × 5 = ___	9 × 6 = ___	9 × 7 = ___	9 × 8 = ___	9 × 9 = ___

Name _____ Date _____

86–87 Write each product.

 ROW 1

1	6	5	7	9	8	5	4
×7	×5	×9	×6	×9	×4	×2	×3

 ROW 2

7	2	9	8	8	9	6	8
×9	×2	×3	×6	×5	×2	×6	×2

 ROW 3

1	9	3	2	7	3	6	2
×6	×4	×7	×5	×4	×9	×1	×9

 ROW 4

2	9	3	4	4	6	8	5
×0	×7	×8	×6	×1	×4	×3	×7

 ROW 5

2	1	7	4	9	7	4	6
×3	×4	×7	×0	×6	×3	×4	×7

 ROW 6

4	4	8	2	3	4	9	6
×9	×2	×7	×4	×6	×8	×5	×8

 ROW 7

0	2	6	5	3	6	7	7
×7	×1	×3	×6	×1	×9	×8	×2

ROW 8

7	5	9	2	6	8	5	4
×1	×3	×8	×6	×2	×9	×8	×5

Name _____ Date _____

Your teacher will tell you what factor to write in the top row of each
table. Multiply each number in the first column by that factor.

×	
4	
7	
2	
5	
9	
1	
6	
0	
8	
6	
3	
5	

×	
6	
1	
8	
0	
3	
8	
5	
9	
7	
2	
3	
4	

×	
2	
6	
3	
8	
6	
0	
7	
5	
1	
9	
4	
7	

Name _____ Date _____

86–87 Write the quotient. Cross out the facts you know.
Keep track of facts you need to study.

÷ 0	÷ 1	÷ 2	÷ 3	÷ 4
	$1 \div 1 =$ ___	$2 \div 2 =$ ___	$3 \div 3 =$ ___	$4 \div 4 =$ ___
	$2 \div 1 =$ ___	$4 \div 2 =$ ___	$6 \div 3 =$ ___	$8 \div 4 =$ ___
	$3 \div 1 =$ ___	$6 \div 2 =$ ___	$9 \div 3 =$ ___	$12 \div 4 =$ ___
You CANNOT divide a number by zero.	$4 \div 1 =$ ___	$8 \div 2 =$ ___	$12 \div 3 =$ ___	$16 \div 4 =$ ___
	$5 \div 1 =$ ___	$10 \div 2 =$ ___	$15 \div 3 =$ ___	$20 \div 4 =$ ___
	$6 \div 1 =$ ___	$12 \div 2 =$ ___	$18 \div 3 =$ ___	$24 \div 4 =$ ___
	$7 \div 1 =$ ___	$14 \div 2 =$ ___	$21 \div 3 =$ ___	$28 \div 4 =$ ___
	$8 \div 1 =$ ___	$16 \div 2 =$ ___	$24 \div 3 =$ ___	$32 \div 4 =$ ___
	$9 \div 1 =$ ___	$18 \div 2 =$ ___	$27 \div 3 =$ ___	$36 \div 4 =$ ___
	$10 \div 1 =$ ___	$20 \div 2 =$ ___	$30 \div 3 =$ ___	$40 \div 4 =$ ___

÷ 5	÷ 6	÷ 7	÷ 8	÷ 9
$5 \div 5 =$ ___	$6 \div 6 =$ ___	$7 \div 7 =$ ___	$8 \div 8 =$ ___	$9 \div 9 =$ ___
$10 \div 5 =$ ___	$12 \div 6 =$ ___	$14 \div 7 =$ ___	$16 \div 8 =$ ___	$18 \div 9 =$ ___
$15 \div 5 =$ ___	$18 \div 6 =$ ___	$21 \div 7 =$ ___	$24 \div 8 =$ ___	$27 \div 9 =$ ___
$20 \div 5 =$ ___	$24 \div 6 =$ ___	$28 \div 7 =$ ___	$32 \div 8 =$ ___	$36 \div 9 =$ ___
$25 \div 5 =$ ___	$30 \div 6 =$ ___	$35 \div 7 =$ ___	$40 \div 8 =$ ___	$45 \div 9 =$ ___
$30 \div 5 =$ ___	$36 \div 6 =$ ___	$42 \div 7 =$ ___	$48 \div 8 =$ ___	$54 \div 9 =$ ___
$35 \div 5 =$ ___	$42 \div 6 =$ ___	$49 \div 7 =$ ___	$56 \div 8 =$ ___	$63 \div 9 =$ ___
$40 \div 5 =$ ___	$48 \div 6 =$ ___	$56 \div 7 =$ ___	$64 \div 8 =$ ___	$72 \div 9 =$ ___
$45 \div 5 =$ ___	$54 \div 6 =$ ___	$63 \div 7 =$ ___	$72 \div 8 =$ ___	$81 \div 9 =$ ___
$50 \div 5 =$ ___	$60 \div 6 =$ ___	$70 \div 7 =$ ___	$80 \div 8 =$ ___	$90 \div 9 =$ ___

Name _____ Date _____

Find each quotient.

86–87

ROW 1 $8\overline{)40}$ $9\overline{)63}$ $2\overline{)6}$ $7\overline{)49}$ $3\overline{)27}$ $8\overline{)32}$ $3\overline{)21}$ $4\overline{)4}$

ROW 2 $8\overline{)64}$ $9\overline{)27}$ $5\overline{)10}$ $8\overline{)24}$ $4\overline{)28}$ $5\overline{)5}$ $3\overline{)6}$ $6\overline{)42}$

ROW 3 $2\overline{)18}$ $6\overline{)36}$ $7\overline{)7}$ $5\overline{)15}$ $4\overline{)12}$ $8\overline{)72}$ $4\overline{)36}$ $3\overline{)3}$

ROW 4 $3\overline{)24}$ $6\overline{)54}$ $4\overline{)16}$ $6\overline{)18}$ $3\overline{)9}$ $5\overline{)45}$ $7\overline{)14}$ $6\overline{)30}$

ROW 5 $8\overline{)16}$ $3\overline{)12}$ $6\overline{)24}$ $8\overline{)56}$ $4\overline{)20}$ $9\overline{)18}$ $9\overline{)45}$ $7\overline{)21}$

ROW 6 $4\overline{)8}$ $7\overline{)63}$ $5\overline{)40}$ $9\overline{)81}$ $3\overline{)15}$ $7\overline{)56}$ $7\overline{)35}$ $9\overline{)54}$

ROW 7 $5\overline{)25}$ $4\overline{)24}$ $6\overline{)48}$ $5\overline{)35}$ $6\overline{)6}$ $8\overline{)48}$ $7\overline{)42}$ $9\overline{)36}$

ROW 8 $2\overline{)8}$ $7\overline{)28}$ $5\overline{)30}$ $9\overline{)72}$ $5\overline{)20}$ $4\overline{)32}$ $3\overline{)18}$ $9\overline{)9}$

Math to Know

Name _____ Date _____

Mastering Basic Facts

Write the sum, difference, product, or quotient.

1. $14 - 8 =$ _____   **2.** $64 \div 8 =$ _____   **3.** $27 \div 9 =$ _____

4. $10 \div 5 =$ _____   **5.** $4 \times 9 =$ _____   **6.** $8 - 6 =$ _____

7. $17 - 9 =$ _____   **8.** $8 \times 7 =$ _____   **9.** $35 \div 7 =$ _____

10. $12 - 8 =$ _____   **11.** $9 - 2 =$ _____   **12.** $4 \times 8 =$ _____

13. $56 \div 7 =$ _____   **14.** $3 + 6 =$ _____   **15.** $15 \div 3 =$ _____

16. $12 - 9 =$ _____   **17.** $7 + 5 =$ _____   **18.** $8 \times 6 =$ _____

19. $6 \div 2 =$ _____   **20.** $81 \div 9 =$ _____   **21.** $49 \div 7 =$ _____

22. $10 - 5 =$ _____   **23.** $7 - 2 =$ _____   **24.** $9 \times 6 =$ _____

25. $5 + 9 =$ _____   **26.** $9 - 3 =$ _____   **27.** $36 \div 9 =$ _____

28. $3 \times 6 =$ _____   **29.** $8 \div 8 =$ _____   **30.** $12 \div 6 =$ _____

31. $7 \times 9 =$ _____   **32.** $16 - 9 =$ _____   **33.** $8 \div 4 =$ _____

34. $6 + 4 =$ _____   **35.** $15 - 6 =$ _____   **36.** $9 \times 3 =$ _____

37. $5 - 4 =$ _____   **38.** $9 \times 5 =$ _____   **39.** $63 \div 7 =$ _____

40. $14 - 7 =$ _____   **41.** $3 + 8 =$ _____   **42.** $24 \div 8 =$ _____

43. $6 \times 7 =$ _____   **44.** $12 - 3 =$ _____   **45.** $40 \div 5 =$ _____

46. $7 \times 3 =$ _____   **47.** $9 - 6 =$ _____   **48.** $4 \times 4 =$ _____

PRACTICE ANSWERS
Page 45 *

Add 0: 0 1 2 3 4 5 6 7 8 9
Add 1: 1 2 3 4 5 6 7 8 9 10
Add 2: 2 3 4 5 6 7 8 9 10 11
Add 3: 3 4 5 6 7 8 10 11 12
Add 4: 4 5 6 7 8 9 10 11 12 13
Add 5: 5 6 7 8 9 10 11 12 13 14
Add 6: 6 7 8 9 10 11 12 13 14 15
Add 7: 7 8 9 10 11 12 13 14 15 16
Add 8: 8 9 10 11 12 13 14 15 16 17
Add 9: 9 10 11 12 13 14 15 16 17 18

Page 46

Row 1: 12 5 9 14 12 7 13 7
Row 2: 8 11 16 8 11 8 13 9
Row 3: 10 10 10 11 14 8 11 9
Row 4: 12 14 9 7 13 13 14 15
Row 5: 18 10 11 9 15 16 17 10
Row 6: 11 11 15 16 8 12 8 15
Row 7: 12 10 13 6 6 11 7 13

Page 47 *

The answers in each column are the same:
0 1 2 3 4 5 6 7 8 9

Page 48

Row 1: 9 3 8 6 4 1 8 1
Row 2: 4 2 0 9 4 3 0 5
Row 3: 6 5 6 0 1 3 0 8
Row 4: 5 4 6 9 4 4 8 7
Row 5: 9 7 1 7 3 8 5 2
Row 6: 0 7 2 5 4 8 8 1
Row 7: 5 9 0 3 8 7 7 6
Row 8: 3 0 2 9 9 6 2 7

Page 49 *

× 0: 0 0 0 0 0 0 0 0 0 0
× 1: 0 1 2 3 4 5 6 7 8 9
× 2: 0 2 4 6 8 10 12 14 16 18
× 3: 0 3 6 9 12 15 18 21 24 27
× 4: 0 4 8 12 16 20 24 28 32 36
× 5: 0 5 10 15 20 25 30 35 40 45
× 6: 0 6 12 18 24 30 36 42 48 54
× 7: 0 7 14 21 28 35 42 49 56 63
× 8: 0 8 16 24 32 40 48 56 64 72
× 9: 0 9 18 27 36 45 54 63 72 81

Page 50

Row 1: 7 30 45 42 81 32 10 12
Row 2: 63 4 27 48 40 18 36 16
Row 3: 6 36 21 10 28 27 6 18
Row 4: 0 63 24 24 4 24 24 35
Row 5: 6 4 49 0 54 21 16 42
Row 6: 36 8 56 8 18 32 45 48
Row 7: 0 2 18 30 3 54 56 14
Row 8: 7 15 72 12 12 72 40 20

Page 51

Answers will vary according to the factor chosen.

Page 52 *

The answers in each column are the same:
1 2 3 4 5 6 7 8 9 10

Page 53

Row 1: 5 7 3 7 9 4 7 1
Row 2: 8 3 2 3 7 1 2 7
Row 3: 9 6 1 3 3 9 9 1
Row 4: 8 9 4 3 3 9 2 5
Row 5: 2 4 4 7 5 2 5 3
Row 6: 2 9 8 9 5 8 5 6
Row 7: 5 6 8 7 1 6 6 4
Row 8: 4 4 6 8 4 8 6 1

TEST PREP ANSWERS
Page 54

1. 6	**2.** 8	**3.** 3
4. 2	**5.** 36	**6.** 2
7. 8	**8.** 56	**9.** 5
10. 4	**11.** 7	**12.** 32
13. 8	**14.** 9	**15.** 5
16. 3	**17.** 12	**18.** 48
19. 3	**20.** 9	**21.** 7
22. 5	**23.** 5	**24.** 54
25. 14	**26.** 6	**27.** 4
28. 18	**29.** 1	**30.** 2
31. 63	**32.** 7	**33.** 2
34. 10	**35.** 9	**36.** 27
37. 1	**38.** 45	**39.** 9
40. 7	**41.** 11	**42.** 3
43. 42	**44.** 9	**45.** 8
46. 21	**47.** 3	**48.** 16

* These pages are designed so that students can cross off facts they know. When the facts are in order like on these pages, it may seem like a simple task to write the answers, particularly for the subtraction and division facts. This simplicity may help students observe some patterns that will help them master the facts. However, the main purpose of these pages is a record-keeping form so that students can cross off the facts that they find "easy." That way, they will be less overwhelmed with the memorization task. They can then focus only on those facts they need to work on.

Name _____ Date _____

Factors and Multiples

88–89

1. Write three factor pairs of 18.

    ____ × ____ = 18   ____ × ____ = 18   ____ × ____ = 18

2. Write three factor pairs of 24.

    ____ × ____ = 24   ____ × ____ = 24   ____ × ____ = 24

3. Write three factor pairs of 16.

    ____ × ____ = 16   ____ × ____ = 16   ____ × ____ = 16

90

4. Write the first six multiples of 4 starting with 4.

    _____   _____   _____   _____   _____   _____

91

5. Write the first six even numbers.

    _____   _____   _____   _____   _____   _____

6. Write the first six odd numbers.

    _____   _____   _____   _____   _____   _____

92–93

Write the factors. Identify the number as prime or composite.

7. Factors of 7: _____ prime   composite

8. Factors of 8: _____ prime   composite

9. Factors of 9: _____ prime   composite

10. Factors of 10: _____ prime   composite

11. Factors of 11: _____ prime   composite

Name _____ Date _____

12. List the factors of 12 and 18. Circle the common factors.   `94`

12 _____

18 _____

13. What is the greatest common factor of 12 and 18? _____

14. List the first eight multiples of 2 and 3. Circle the common multiples.   `95`

2 _____ _____ _____ _____ _____ _____ _____ _____

3 _____ _____ _____ _____ _____ _____ _____ _____

15. What is the least common multiple of 2 and 3? _____

16. Write the prime factors of 24. Use a factor tree.   `96–97`

6 × 6 can be written as with 6 as the base and 2 as an exponent → 6^2.   `98–99`
6^2 in standard form is 36.

Write each using exponents. Then write the standard form.

	Using Exponents	Standard Form
17. 5×5	_____	_____
18. $3 \times 3 \times 3$	_____	_____
19. $2 \times 2 \times 2 \times 2$	_____	_____

Name _____ Date _____

Factors and Multiples

Directions: Circle the letter beside the correct answer.

1. Which is a factor pair of 6?

   A. 1 and 3

   B. 2 and 3

   C. 1 and 5

2. Which are multiples of 4?

   A. 4, 6, 8

   B. 4, 8, 12

   C. 4, 8, 10

3. Which are even numbers?

   A. 5, 25, 33

   B. 6, 10, 34

   C. 21, 47, 9

4. Which is a prime number?

   A. 3

   B. 6

   C. 9

5. Which is a composite number?

   A. 2

   B. 10

   C. 11

6. Which is a common factor of 6 and 15?

   A. 2

   B. 3

   C. 5

7. Which is a common multiple of 4 and 5?

   A. 12

   B. 20

   C. 25

8. Which number does 3^2 equal?

   A. 5

   B. 6

   C. 9

PRACTICE ANSWERS
Page 56

1. 1, 18
2. Any 3 of the following:
   1, 24
   2, 12
   3, 8
   4, 6
3. 1, 16
   2, 8
   4, 4
4. 4, 8, 12, 16, 20, 24
5. 2, 4, 6, 8, 10, 12
6. 1, 3, 5, 7, 9, 11
7. 1, 7, prime
8. 1, 2, 4, 8, composite
9. 1, 3, 9 composite
10. 1, 2, 5, 10, composite
11. 1, 11, prime

Page 57

12. 12: ①, ②, ③, 4, ⑥, 12
    18: ①, ②, ③, ⑥, 9, 18
13. 6
14. 2: 2, 4, ⑥, 8, 10, ⑫,
    14, 16
    3: 3, ⑥, 9, ⑫, 15, 18,
    21, 24
15. 6
16. The prime factors of 24 are: 2, 2, 2, and 3.

    One possible factor tree is:

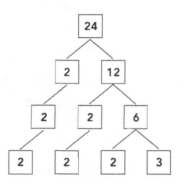

17. 5^2, 25
18. 3^3, 27
19. 2^4, 16

TEST PREP ANSWERS
Page 58

1. B
2. B
3. B
4. A
5. B
6. B
7. B
8. C

Fair Prices

OBJECTIVES
- Use basic facts to solve word problems
- Use a picture to find data needed to solve problems

MATERIALS
- none

TIME
- 20–30 minutes

TEACHER NOTES
- If students need practice determining what operation to use, pose some sample problems. Have them tell whether they would add, subtract, multiply, or divide to solve the problem.

    Sample problems:

    How much money would you need for your admission and a hot dog?

    How much more does a hot dog cost than a lemonade?

    How much will 3 rides cost?

    How many ride tickets can you buy with $8?

EXTENSIONS
- Encourage students to find as many different answers as possible for problem 7.
- Have students make up their own word problems using the data in the chart.
- Have students redo the chart, changing the numbers. Then have them make up other problems and exchange problems with other students.

ANSWERS
1. $9 + $5 = $14
2. $9 − $5 = $4
3. 2 × $9 = $18
4. 6 × $5 = $30
5. 8 × $2 = $16
6. $12 ÷ $2 = 6
7. Possible solution: 3 hot dogs and 1 lemonade. Split the third hot dog and the lemonade.
   Another solution: 2 hot dogs and 4 lemonades. Each can have one hot dog and two lemonades.

Name _____ Date _____

Addition Concepts
and Facts
pages 34-45

Subtraction Concepts
and Facts
pages 46-59

Multiplication
Concepts and Facts
pages 60-73

Division Concepts
and Facts
pages 74-85

Fair

Admission tickets
Adults	$9
Children	$5
Rides	$2
Hot dogs	$3
Lemonade	$1

Fair Prices

Write a number sentence to show how you solve each of problems 1-6.

1. How much will it cost to buy admission tickets for one adult and one child?

_____

2. How much more does an adult ticket cost than a child's ticket?

_____

3. How much will 2 adult tickets cost?

_____

4. How much will 6 childrens' tickets cost?

_____

5. What will you have to pay to go on 8 rides?

_____

6. Your brother has $12 for rides. How many rides can he take?

_____

7. You have $10 to share with your brother for food and drinks. Explain how you will spend the $10. Use all of the money.

_____

_____

Name _____ Date _____

Mental Math—Addition

103   Add. Use mental math.

1. $5 + 1 + 5 + 7 =$ _____   **2.** $3 + 2 + 1 + 8 + 4 =$ _____

3. $6 + 5 + 3 + 4 + 5 =$ _____   **4.** $8 + 6 + 5 + 2 + 2 =$ _____

5. $5 + 5 + 2 + 5 + 3 =$ _____   **6.** $1 + 6 + 4 + 5 + 9 =$ _____

104   **50 + 80 = ■**   **500 + 800 = ■**

5 tens + 8 tens = 13 tens   **5 hundreds + 8 hundreds = 13 hundreds**

50 + 80 = 130   **500 + 800 = 1300**

Add. Use mental math.

7. $50 + 90 =$ _____   **8.** $70 + 40 =$ _____   **9.** $500 + 600 =$ _____

10. $40 + 300 + 20 + 500 =$ _____   **11.** $100 + 400 + 70 + 60 =$ _____

12. $80 + 500 + 600 + 20 =$ _____   **13.** $400 + 60 + 300 + 80 =$ _____

105   **56 + 32 = ■**

Start with the first addend. Add the tens of the second addend and then the ones.

For 56 + 32, think 56 + 30 = 86, then 86 + 2 = 88.
56 + 32 = 88

Add on tens and then ones.

14. $46 + 23 =$ _____   **15.** $65 + 15 =$ _____   **16.** $37 + 44 =$ _____

17. $54 + 35 =$ _____   **18.** $38 + 24 =$ _____   **19.** $58 + 67 =$ _____

20. $75 + 15 =$ _____   **21.** $47 + 45 =$ _____   **22.** $36 + 25 =$ _____

Name _____ Date _____

Add. Try breaking up one or more addends.

106–107

23. $34 + 25 =$ _____　　　　**24.** $57 + 24 =$ _____

25. $62 + 30 + 16 + 45 =$ _____　　**26.** $76 + 15 + 48 + 32 =$ _____

Add. Look for compatible numbers.

108

27. $10 + 40 + 20 + 60 =$ _____　　**28.** $300 + 400 + 700 + 100 =$ _____

29. $70 + 60 + 30 + 20 =$ _____　　**30.** $20 + 50 + 80 + 40 =$ _____

31. $150 + 300 + 250 + 40 =$ _____　　**32.** $130 + 240 + 800 + 270 =$ _____

Add.

109

33.　　50　　　　**34.**　　75　　　　**35.**　　125
　　　　$+50$　　　　　　　　$+25$　　　　　　　　$+75$

You can break up numbers to make compatible numbers.

$$
\begin{aligned}
52 &= 50 + 2 \\
+54 &= \underline{50 + 4} \\
&\ 100 + 6 = 106
\end{aligned}
$$

Add. Try making compatible numbers.

36.　　26　　　　**37.**　　56　　　　**38.**　　53
　　　　$+78$　　　　　　　　$+51$　　　　　　　　$+55$

39.　　76　　　　**40.**　　28　　　　**41.**　　79
　　　　$+27$　　　　　　　　$+77$　　　　　　　　$+26$

Name _____ Date _____

110 **To add 9, add 10 and then subtract 1.**

Add.

42. $15 + 9 =$ _____   **43.** $38 + 9 =$ _____   **44.** $53 + 9 =$ _____

45. $9 + 26 =$ _____   **46.** $72 + 9 =$ _____   **47.** $9 + 49 =$ _____

To add 99, add 100 and subtract 1.

Add.

48. $\begin{array}{r} 153 \\ +99 \\ \hline \end{array}$ **49.** $\begin{array}{r} 467 \\ +99 \\ \hline \end{array}$ **50.** $\begin{array}{r} 538 \\ +99 \\ \hline \end{array}$

51. $\begin{array}{r} 240 \\ +99 \\ \hline \end{array}$ **52.** $\begin{array}{r} 153 \\ +99 \\ \hline \end{array}$ **53.** $\begin{array}{r} 467 \\ +99 \\ \hline \end{array}$

111 **You can use the "give-and-take method" to add. You can start by making one addend a multiple of 10.**

$$\begin{array}{r} 67 + 3 = \quad 70 \\ +34 - 3 = +\,31 \\ \hline 101 \end{array}$$

Add.

54. $\begin{array}{r} 37 \\ +25 \\ \hline \end{array}$ **55.** $\begin{array}{r} 68 \\ +17 \\ \hline \end{array}$ **56.** $\begin{array}{r} 43 \\ +27 \\ \hline \end{array}$

57. $\begin{array}{r} 22 \\ +18 \\ \hline \end{array}$ **58.** $\begin{array}{r} 87 \\ +15 \\ \hline \end{array}$ **59.** $\begin{array}{r} 36 \\ +44 \\ \hline \end{array}$

Name _____ Date _____

Mental Math—Subtraction

Think about the first number. Subtract the tens and then the ones.   `112`

Subtract.

1. $67 - 23 =$ _____   **2.** $84 - 32 =$ _____   **3.** $59 - 46 =$ _____

4. $54 - 16 =$ _____   **5.** $72 - 56 =$ _____   **6.** $48 - 25 =$ _____

To subtract 9, just subtract 10 and add 1.   `113`
To subtract 99, just subtract 100 and add 1.

7. $46 - 9 =$ _____   **8.** $34 - 9 =$ _____   **9.** $72 - 9 =$ _____

10. $327 - 99 =$ _____   **11.** $265 - 99 =$ _____   **12.** $642 - 99 =$ _____

You can break up a number to make subtraction easier.   `114-115`

13. $64 - 38 =$ _____   **14.** $52 - 27 =$ _____
(Think of 38 as $34 + 4$.)   (Think of 27 as $22 + 5$.)

15. $33 - 17 =$ _____   **16.** $81 - 55 =$ _____   **17.** $46 - 14 =$ _____

If you add the same amount to both the number that you're subtracting   `116`
and to the number you started with, the difference will be the same.

$75 - 27 =$ ■

Add 3 to each number. → $78 - 30 = 48$

18. $56 - 29 =$ _____   **19.** $74 - 49 =$ _____   **20.** $86 - 39 =$ _____

21. $80 - 50 =$ _____   **22.** $600 - 200 =$ _____   **23.** $400 - 80 =$ _____   `117`

24. $600 - 50 =$ _____   **25.** $1000 - 300 =$ _____   **26.** $3000 - 800 =$ _____

Name _____ Date _____

Mental Math—Multiplication

118 Multiply.

1. $3 \times 25 =$ _____   **2.** $4 \times 25 =$ _____   **3.** $7 \times 25 =$ _____

4. $5 \times 100 =$ _____   **5.** $4 \times 200 =$ _____   **6.** $6 \times 50 =$ _____

119 Multiply.

7. $2 \times 12 =$ _____   **8.** $4 \times 12 =$ _____   **9.** $8 \times 12 =$ _____

10. $2 \times 23 =$ _____   **11.** $4 \times 23 =$ _____   **12.** $8 \times 23 =$ _____

13. $4 \times 34 =$ _____   **14.** $8 \times 21 =$ _____   **15.** $4 \times 22 =$ _____

120 Multiply each number by 10, 100, and 1000.

	Number	Number $\times$ 10	Number $\times$ 100	Number $\times$ 1000
16.	7			
17.	74			
18.	748			

19. $6 \times 100 =$ _____   **20.** $7 \times 1000 =$ _____   **21.** $45 \times 100 =$ _____

22. $581 \times 10 =$ _____   **23.** $395 \times 100 =$ _____   **24.** $206 \times 1000 =$ _____

121 Multiply each number by 80, 800, and 8000.

	Number	Number $\times$ 80	Number $\times$ 800	Number $\times$ 8000
25.	4			
26.	40			
27.	400			

28. $6 \times 70 =$ _____   **29.** $30 \times 80 =$ _____   **30.** $20 \times 400 =$ _____

31. $800 \times 20 =$ _____   **32.** $40 \times 9000 =$ _____   **33.** $600 \times 3000 =$ _____

Name _____ Date _____

Multiply each number by 10, 5, and 9.

122

	Number	Number × 10	Number × 5	Number × 9
34.	14			
35.	18			
36.	26			

37. $16 \times 10 =$ _____   **38.** $25 \times 10 =$ _____   **39.** $36 \times 10 =$ _____

40. $16 \times 5 =$ _____   **41.** $25 \times 9 =$ _____   **42.** $36 \times 5 =$ _____

Multiply.

123

43. $3 \times 29 =$ _____   Think: $(3 \times 20) + (3 \times 9)$

44. $4 \times 26 =$ _____   **45.** $5 \times 42 =$ _____   **46.** $8 \times 34 =$ _____

Multiply.

124

47. $11 \times 4 =$ _____   **48.** $11 \times 6 =$ _____   **49.** $11 \times 3 =$ _____

50. $11 \times 9 =$ _____   **51.** $11 \times 7 =$ _____   **52.** $11 \times 5 =$ _____

53. $12 \times 3 =$ _____   **54.** $12 \times 5 =$ _____   **55.** $12 \times 9 =$ _____

56. $12 \times 7 =$ _____   **57.** $12 \times 4 =$ _____   **58.** $12 \times 6 =$ _____

Multiply.

125

59. $2 \times 5 \times 7 =$ _____   **60.** $4 \times 5 \times 7 \times 2 =$ _____

61. $25 \times 3 \times 4 =$ _____   **62.** $3 \times 50 \times 6 \times 2 =$ _____

63. $7 \times 20 \times 3 \times 5 =$ _____   **64.** $3 \times 5 \times 8 \times 2 =$ _____

65. $5 \times 4 \times 7 \times 5 =$ _____   **66.** $5 \times 4 \times 3 \times 5 =$ _____

67. $2 \times 25 \times 9 \times 2 =$ _____   **68.** $60 \times 2 \times 5 \times 7 =$ _____

Name _____ Date _____

Mental Math—Division

126 Divide.

1. $46 \div 2 =$ _____   2. $28 \div 2 =$ _____   3. $64 \div 2 =$ _____

4. $24 \div 2 =$ _____   5. $34 \div 2 =$ _____   6. $42 \div 2 =$ _____

7. $88 \div 2 =$ _____   8. $72 \div 2 =$ _____   9. $56 \div 2 =$ _____

10. $468 \div 2 =$ _____   11. $624 \div 2 =$ _____   12. $128 \div 2 =$ _____

13. $40 \div 10 =$ _____   14. $400 \div 10 =$ _____   15. $450 \div 10 =$ _____

16. $60 \div 10 =$ _____   17. $720 \div 10 =$ _____   18. $890 \div 10 =$ _____

127 Divide.

19. $8 \div 4 =$ _____   20. $80 \div 4 =$ _____   21. $800 \div 4 =$ _____

22. $12 \div 3 =$ _____   23. $120 \div 3 =$ _____   24. $1200 \div 3 =$ _____

25. $24 \div 8 =$ _____   26. $2400 \div 8 =$ _____   27. $24{,}000 \div 8 =$ _____

28. $560 \div 7 =$ _____   29. $45{,}000 \div 9 =$ _____   30. $2800 \div 4 =$ _____

31. $810 \div 9 =$ _____   32. $4200 \div 7 =$ _____   33. $8100 \div 9 =$ _____

34. $720 \div 8 =$ _____   35. $4800 \div 8 =$ _____   36. $120 \div 2 =$ _____

37. $1000 \div 2 =$ _____   38. $3600 \div 6 =$ _____   39. $5400 \div 6 =$ _____

40. $420 \div 6 =$ _____   41. $1200 \div 4 =$ _____   42. $720 \div 9 =$ _____

43. $8100 \div 9 =$ _____   44. $300 \div 5 =$ _____   45. $4200 \div 7 =$ _____

46. $90 \div 3 =$ _____   47. $420 \div 6 =$ _____   48. $400 \div 8 =$ _____

49. $1200 \div 4 =$ _____   50. $900 \div 9 =$ _____   51. $1600 \div 4 =$ _____

52. $700 \div 7 =$ _____   53. $4000 \div 5 =$ _____   54. $600 \div 3 =$ _____

Name _____ Date _____

Mental Math

Directions: Use mental math to solve each problem. Write your answers.

34¢   59¢   76¢   25¢

How much will it cost to buy

1.  _____

2.  _____

3.  _____

4.  _____

5.  _____

6.  _____

Solve.

7. You buy 1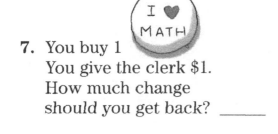
You give the clerk $1.
How much change
should you get back? _____

8. There are 800 erasers. You will
share them evenly among 4
schools. How many erasers
will each school get?

_____

PRACTICE ANSWERS
Page 62
1. 18   2. 18
3. 23   4. 23
5. 20   6. 25
7. 140   8. 110   9. 1100
10. 860   11. 630
12. 1200   13. 840
14. 69   15. 80   16. 81
17. 89   18. 62   19. 125
20. 90   21. 92   22. 61

Page 63
23. 59   24. 81
25. 153   26. 171
27. 130   28. 1500
29. 180   30. 190
31. 740   32. 1440
33. 100   34. 100   35. 200
36. 104   37. 107   38. 108
39. 103   40. 105   41. 105

Page 64
42. 24   43. 47   44. 62
45. 35   46. 81   47. 58
48. 252   49. 566   50. 637
51. 339   52. 252   53. 566
54. 62   55. 85   56. 70
57. 40   58. 102   59. 80

Page 65
1. 44   2. 52   3. 13
4. 38   5. 16   6. 23
7. 37   8. 25   9. 61
10. 228   11. 166   12. 543
13. 26   14. 25
15. 16   16. 26   17. 32
18. 27   19. 25   20. 47
21. 30   22. 400   23. 320
24. 550   25. 700   26. 2200

Page 66
1. 75   2. 100   3. 175
4. 500   5. 800   6. 300
7. 24   8. 48   9. 96
10. 46   11. 92   12. 184
13. 136   14. 168   15. 88
16. 70; 700; 7000
17. 740; 7400; 74,000
18. 7480; 74,800; 748,000
19. 600   20. 7000   21. 4500
22. 5810   23. 39,500
24. 206,000
25. 320; 3200; 32,000
26. 3200; 32,000 320,000
27. 32,000; 320,000;
    3,200,000
28. 420   29. 2400   30. 8000
31. 16,000
32. 360,000
33. 1,800,000

Page 67
34. 140; 70; 126
35. 180; 90; 162
36. 260; 130; 234
37. 160   38. 250   39. 360
40. 80   41. 225   42. 180
43. 87
44. 104   45. 210   46. 272
47. 44   48. 66   49. 33
50. 99   51. 77   52. 55
53. 36   54. 60   55. 108
56. 84   57. 48   58. 72
59. 70   60. 2800
61. 300   62. 1800
63. 2100   64. 240
65. 700   66. 300
67. 900   68. 4200

Page 68
1. 23   2. 14   3. 32
4. 12   5. 17   6. 21
7. 44   8. 36   9. 28
10. 234   11. 312   12. 64
13. 4   14. 40   15. 45
16. 6   17. 72   18. 89
19. 2   20. 20   21. 200
22. 4   23. 40   24. 400
25. 3   26. 300   27. 3000
28. 80   29. 5000   30. 700
31. 90   32. 600   33. 900
34. 90   35. 600   36. 60
37. 500   38. 600   39. 900
40. 70   41. 300   42. 80
43. 900   44. 60   45. 600
46. 30   47. 70   48. 50
49. 300   50. 100   51. 400
52. 100   53. 800   54. 200

TEST PREP ANSWERS
Page 69
1. 93¢
2. $1.01
3. $1.50
4. $3.40
5. $1.18
6. $4.72
7. 66¢
8. 200 erasers

Name _____ Date _____

Estimation

Round each number to the nearest ten.   128–130

1. 42 _____   **2.** 28 _____   **3.** 35 _____   **4.** 59 _____

5. 741 _____   **6.** 976 _____   **7.** 2736 _____   **8.** 4782 _____

Round each number to the nearest hundred.

9. 328 _____   **10.** 274 _____   **11.** 450 _____   **12.** 1293 _____

Round each number to the nearest thousand.

13. 7241 _____   **14.** 3098 _____   **15.** 4620 _____   **16.** 1903 _____

Round each number to the nearest whole number.   131

17. 3.8 _____   **18.** 7.3 _____   **19.** 1.5 _____

20. 28.6 _____   **21.** 72.9 _____   **22.** 0.8 _____

Round each number to the nearest tenth.

23. 7.48 _____   **24.** 5.84 _____   **25.** 12.09 _____   **26.** 1.65 _____

Estimate each sum or difference by rounding each number to   132–13
the nearest ten. Show the rounded numbers you used.

27. 48 + 23   **28.** 39 + 55   **29.** 138 + 41

_____   _____   _____

30. 68 − 24   **31.** 39 − 12   **32.** 272 − 239

_____   _____   _____

33. $39 + $24   **34.** $76 − $18   **35.** $261 − $215

_____   _____   _____

Math to Know

Name _____ Date _____

132–133 Estimate each sum or difference by rounding each number to the nearest hundred. Show the rounded numbers you used.

36. $328 + 412$ **37.** $584 + 140$ **38.** $402 - 156$

_____   _____   _____

39. $492 + $313   **40.** $861 - $27   **41.** $762 - $190

_____   _____   _____

134–135 Use front-end estimation to estimate the sum or difference. Show the numbers you used.

42. $418 + 266$ **43.** $755 + 119$ **44.** $28 + 619 + 231$

_____   _____   _____

45. $825 - 258$ **46.** $4.52 + $3.05   **47.** $7.86 - $3.27

_____   _____   _____

136–137 Estimate each product by rounding *one* factor. Show the numbers you used.

48. 42×5 **49.** 68×2 **50.** 718×6 **51.** 494×3

_____   _____   _____   _____

Estimate the product by rounding *each* factor. Show the numbers you used.

52. 32×49 **53.** 88×57 **54.** 593×24 **55.** 789×36

_____   _____   _____   _____

Use front-end estimation to estimate each product. Show the numbers you used.

56. $3 \times $4.28   **57.** $7 \times $8.30   **58.** $6 \times $7.19

_____   _____   _____

59. $5 \times $9.44   **60.** $4 \times $7.05   **61.** $8 \times $3.28

_____   _____   _____

Name _____ Date _____

Estimate the product by looking for compatible numbers.
Write the numbers you used.

62. 9×27 **63.** 5×19 **64.** 4×247

_____   _____   _____

65. 11×721 **66.** 8×48 **67.** 49×188

_____   _____   _____

Estimate the quotient by looking for compatible numbers.
Write the numbers you used.

Example: $365 \div 6 \to 360 \div 6 = 60$

68. $573 \div 8$ **69.** $462 \div 9$ **70.** $735 \div 8$

_____   _____   _____

71. $223 \div 3$ **72.** $439 \div 7$ **73.** $294 \div 4$

_____   _____   _____

Write the number of digits that will be in each quotient.

74. $7\overline{)522}$ _____   **75.** $3\overline{)732}$ _____   **76.** $5\overline{)1374}$ _____   **77.** $4\overline{)9253}$ _____

Use benchmark fractions of 0, $\frac{1}{2}$, or 1 to estimate the sum or difference.
Show the benchmark fractions you used.

78. $\frac{5}{8} + \frac{7}{8}$ **79.** $\frac{1}{12} + \frac{7}{12}$ **80.** $\frac{3}{16} + \frac{15}{16}$

_____   _____   _____

81. $\frac{5}{12} + \frac{3}{8}$ **82.** $\frac{9}{16} - \frac{1}{12}$ **83.** $\frac{11}{12} - \frac{7}{16}$

_____   _____   _____

84. $\frac{14}{16} - \frac{2}{12}$ **85.** $2\frac{7}{16} + 1\frac{1}{8}$ **86.** $5\frac{3}{16} - 2\frac{7}{8}$

_____   _____   _____

Name _____ Date _____

Estimation

Directions: Fill in the circle beside the letter of the best answer.

1. 976 rounded to the nearest ten is

○ **A.** 970

○ **B.** 980

○ **C.** 1000

2. 6549 rounded to the nearest hundred is

○ **A.** 7000

○ **B.** 6600

○ **C.** 6500

3. Estimate 38 + 53.

○ **A.** 90

○ **B.** 100

○ **C.** 70

4. Estimate 853 − 276 by rounding each number to the nearest hundred.

○ **A.** 600

○ **B.** 500

○ **C.** 400

5. Estimate 4 × 625.

○ **A.** 2000

○ **B.** 2400

○ **C.** 2800

6. How many digits will be in the quotient for 7352 ÷ 8?

○ **A.** 2

○ **B.** 3

○ **C.** 4

7. What basic fact will help you find the quotient for 468 ÷ 5?

○ **A.** 45 ÷ 5 = 9

○ **B.** 42 ÷ 6 = 7

○ **C.** 48 ÷ 8 = 6

8. Estimate $\frac{5}{9} + \frac{15}{16}$.

○ **A.** $\frac{1}{2}$

○ **B.** 1

○ **C.** $1\frac{1}{2}$

PRACTICE ANSWERS
Page 71

1. 40
2. 30
3. 40
4. 60
5. 740
6. 980
7. 2740
8. 4780
9. 300
10. 300
11. 500
12. 1300
13. 7000
14. 3000
15. 5000
16. 2000
17. 4   18. 7   19. 2
20. 29   21. 73   22. 1
23. 7.5
24. 5.8
25. 12.1
26. 1.7
27. $50 + 20 = 70$
28. $40 + 60 = 100$
29. $140 + 40 = 180$
30. $70 - 20 = 50$
31. $40 - 10 = 30$
32. $270 - 240 = 30$
33. $40 + \$20 = \$60$
34. $80 - \$20 = \$60$
35. $260 - \$220 = \$40$

Page 72

36. $300 + 400 = 700$
37. $600 + 100 = 700$
38. $400 - 200 = 200$
39. $\$500 + \$300 = \$800$
40. $\$900 - \$0 = \$900$
41. $\$800 - \$200 = \$600$
42. $400 + 200 = 600$
43. $700 + 100 = 800$
44. $20 + 610 + 230 = 860$
45. $800 - 200 = 600$
46. $\$4 + \$3 = \$7$
47. $\$7 - \$3 = \$4$
48. $40 \times 5 = 200$
49. $70 \times 2 = 140$
50. $700 \times 6 = 4200$
51. $500 \times 3 = 1500$
52. $30 \times 50 = 1500$
53. $90 \times 60 = 5400$
54. $600 \times 20 = 12,000$
55. $800 \times 40 = 3200$
56. $3 \times \$4 = \$12$
57. $7 \times \$8 = \$56$
58. $6 \times \$7 = \$42$
59. $5 \times \$9 = \$45$
60. $4 \times \$7 = \$28$
61. $8 \times \$3 = \$24$

Page 73

62. $9 \times 30 = 270$
    or $10 \times 27 = 270$
63. $5 \times 20 = 100$
64. $4 \times 250 = 1000$
65. $10 \times 700 = 7000$
66. $8 \times 50 = 400$
67. $50 \times 200 = 10,000$
68. $560 \div 8 = 70$
69. $450 \div 9 = 50$
70. $720 \div 8 = 90$
71. $210 \div 3 = 70$
    or $240 \div 3 = 80$
72. $420 \div 7 = 60$
73. $280 \div 4 = 70$
74. 2
75. 3
76. 3
77. 4
78. $\frac{1}{2} + 1 = 1\frac{1}{2}$
79. $0 + \frac{1}{2} = \frac{1}{2}$
80. $0 + 1 = 1$
81. $\frac{1}{2} + \frac{1}{2} = 1$
82. $\frac{1}{2} - 0 = \frac{1}{2}$
83. $1 - \frac{1}{2} = \frac{1}{2}$
84. $1 - 0 = 1$
85. $2\frac{1}{2} + 1 = 3\frac{1}{2}$
86. $5 - 3 = 2$

TEST PREP ANSWERS
Page 74

1. B
2. C
3. A
4. A
5. B
6. B
7. A
8. C

A Body of Numbers

OBJECTIVE
• Solve problems using mental math and estimation

MATERIALS
• clock or stopwatch for measuring one minute

TIME
• 30–40 minutes

TEACHER NOTES
• For problem 5, students will need to take their pulse. Demonstrate how this is done.

• Review the number of hours in a day (24) and the number of minutes in an hour (60).

EXTENSIONS
• Compare students' estimates for problem 3. See how many different methods the members of the class used. Try to have the class come up with all the different methods that could be used. Then have students find the exact answer. Ask: "Which estimate is closest to the actual answer?"

• Have students use the data used on page 77 to make up their own problems similar to those on the page.

• Have students research their own information about the human body or another topic that interests them. Then ask them to use the information they collect to make up problems like the ones on page 77.

ANSWERS
1. 700

2. 12 more teeth

Estimates for problems 3–6 will vary depending on methods used.

3. 150 (350 − 200 = 150)
   or 140 (350 − 210 = 140)
   or 100 (300 − 200)
   or 200 (400 − 200) fewer bones

4. 20 (40 − 20 = 20)
   or 30 (40 − 10) fewer muscles

5. For 70 beats/minute:
   70 × 60 min = 4200 beats/h

   Possible methods to estimate beats/day:

   4200 × 24 h = 100,800 beats/day

   4000 × 20 = 80,000 beats/day

   4000 × 25 = 100,000 beats/day

6. 80 × 20 = 1600
   or 80 × 25 = 2000 gallons

Name _____ Date _____

A Body of Numbers

Mental Subtraction
pages 112-115

Rounding Numbers
pages 128-131

Estimating Sums and Differences
pages 132-135

Estimating Products
pages 136-138

1. The human body has more than 650 muscles. Rounded to the nearest hundred, the human body

   has about _____ muscles.

2. Children have 20 first teeth. Adults have 32 teeth. Use mental math to find how many more teeth

   adults have than children._____

Make an estimate for each situation. Write the numbers you used to make your estimate.

3. A baby's body has 350 bones. An adult's body has 206 bones. That's because bones grow together to make fewer, larger bones as a baby grows up. Estimate how many fewer bones an adult has than a baby has.

   Estimate: _____ The numbers I used: _____

4. It takes around 17 muscles to smile and 43 muscles to frown. Estimate how many fewer muscles it takes to smile than to frown.

   Estimate: _____ The numbers I used: _____

5. Find out how many times your heart beats in one minute. Round that number to the nearest ten. Estimate the number of times your heart beats in one day.

   Estimate: _____ The numbers I used: _____

6. The average heart pumps about 83 gallons of blood in one hour. Estimate the number of gallons of blood the heart pumps in a day.

   Estimate: _____ The numbers I used: _____

Math to Know

Name _____ Date _____

Addition

 146–147 Use the grid to line up the digits by place value. Find the sum.

1. 43 + 35   **2.** 60 + 15   **3.** 503 + 74   **4.** 2375 + 120

    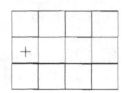 

148–149 Find the sum.

5. 37
    +49

6. 582
    +364

7. 129
    +658

8. 607
    +384

9. 231
    +429

10. 532
    +816

11. 1206
    +2356

12. 508
    +109

13. 4305
    +3498

150–151 **14.** 74   **15.** 76   **16.** 29   **17.** 64   **18.** 48
    +59   +38   +45   +36   +20

19. 148   **20.** 728   **21.** 825   **22.** 603
    +509   +291   + 96   +196

Name _____ Date _____

Check these addition problems by changing the order of the addends. Write the correct answer for each problem.

152

23.
```
   999          573
  +573        +_____
  1462
```
correct answer: _____

24.
```
   506          837
  +837        +_____
  1443
```
correct answer: _____

Arrange the numbers in a column and add.

153

25. 48, 8, 206

26. 1306, 45, 2, 936

Add.

154–157

27.
```
  0.5
 +0.2
```

28.
```
  3.4
 +1.5
```

29.
```
  0.1
 +1.2
```

30.
```
  6.32
 +0.45
```

31.
```
  4.63
 +2.92
```

Arrange the numbers in a column and add.

32. 2.52 + 0.6

33. 3.8 + 4

34. 8.12 + 0.36 + 0.4

35. $2.48 + $3.66

36. $5 + $3.54

37. $2.98 + $3.65

158

Name _____   Date _____

Addition

Directions: Compute. Put a check beside the letter of the correct answer.

1. 34 + 25

   A. 58

   B. 59

   C. 69

2. 48 + 16

   A. 514

   B. 54

   C. 64

3. 159 + 222

   A. 381

   B. 371

   C. 372

4. 574 + 271

   A. 845

   B. 7145

   C. 745

5. 65 + 45

   A. 100

   B. 110

   C. 120

6. 806 + 407

   A. 1213

   B. 123

   C. 1203

7. 634 + 255

   A. 899

   B. 889

   C. 999

8. 35 + 18 + 66 + 45

   A. 174

   B. 156

   C. 164

9. 1.38 + 0.56

   A. 6.98

   B. 1.94

   C. 1.84

10. 2 + 0.6 + 0.58

   A. 2.64

   B. 2.18

   C. 3.18

PRACTICE ANSWERS
Page 78

1. 78
2. 75
3. 577
4. 2495
5. 86
6. 946
7. 787
8. 991
9. 660
10. 1348
11. 3562
12. 617
13. 7803
14. 133
15. 114
16. 74
17. 100
18. 68
19. 657
20. 1019
21. 921
22. 799

Page 79

23.
$$\begin{array}{r} 573 \\ +999 \\ \hline 1572 \end{array}$$
correct answer: 1572

24.
$$\begin{array}{r} 837 \\ +506 \\ \hline 1343 \end{array}$$
correct answer: 1343

25. 262
26. 2289
27. 0.7
28. 4.9
29. 1.3
30. 6.77
31. 7.55
32. 3.12
33. 7.8
34. 8.88
35. $6.14
36. $8.54
37. $6.63

TEST PREP ANSWERS
Page 80

1. B
2. C
3. A
4. A
5. B
6. A
7. B
8. C
9. B
10. C

Name _____ Date _____

Subtraction

159–160 Use the grid to line up the digits by place value. Find the difference.

1. 49 − 37   **2.** 96 − 35   **3.** 769 − 304   **4.** 583 − 42

    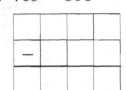 

161 Find the difference.

5.   35   **6.**   86   **7.**   92   **8.**   60
       −17   −67   −58   −34

9.   63   **10.**   86   **11.**   50   **12.**   848
       −46   −49   − 6   −317

162–163 **13.**   254   **14.**   926   **15.**   367   **16.**   674
              −181   −430   −182   −391

17.   540   **18.**   348   **19.**   868   **20.**   268
       −160   −217   −673   − 76

21.   413   **22.**   537   **23.**   725   **24.**   925
       − 82   −184   −183   − 84

Name _____ Date _____

Use addition to check these subtraction answers.
Write the correct answer for each problem.

163

25.   777   594   **26.**   806   361
        −283   + _____   −435   + _____
        ────   ────
        594   361

      correct answer: _____   correct answer: _____

27.   874   216   **28.**   450   246
        −680   + _____   −196   + _____
        ────   ────
        216   246

      correct answer: _____   correct answer: _____

Find the difference.

29.   768   **30.**   250   **31.**   342   **32.**   435
        −299   −181   − 87   −182

33.   204   **34.**   906   **35.**   207   **36.**   504
        −186   −539   −197   − 91

166–167

37.   500   **38.**   4000   **39.**   9014
        −163   − 37   −2675

40.   1007   **41.**   1004   **42.**   2000
        − 547   − 68   − 370

Name _____ Date _____

 Arrange the numbers in a column and subtract.
Be sure to line up the digits by place value.

43. 0.8 − 0.6   **44.** 3.4 − 1.6   **45.** 1.3 − 0.6

46. 2.3 − 0.5   **47.** 9.3 − 8.7   **48.** 7.8 − 1.8

170 **49.** 3.63 − 1.28   **50.** 7.36 − 0.45   **51.** 9.03 − 5.76

171 Use the grid to line up the decimal points and the digits
by place value. If the right side is uneven, tack on zeros.
Find the difference.

52. 4.21 − 0.62   **53.** 5.8 − 4.28   **54.** 8.36 − 0.4

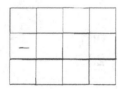   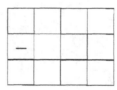   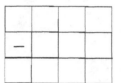

Find the difference.

55. $5.42 − $1.98   **56.** $6 − $2.75   **57.** $9.28 − $3.58

Name _____ Date _____

Subtraction

Directions: Fill in the circle beside the letter of the correct answer.

1. 695
   −473

   ○ **A.** 128
   ○ **B.** 222
   ○ **C.** 258

2. 262
   −190

   ○ **D.** 72
   ○ **E.** 132
   ○ **F.** 178

3. 730
   −468

   ○ **A.** 338
   ○ **B.** 362
   ○ **C.** 262

4. 6002
   −2574

   ○ **D.** 3428
   ○ **E.** 4572
   ○ **F.** 3328

5. 7.45
   −3.29

   ○ **A.** 3.16
   ○ **B.** 4.16
   ○ **C.** 4.24

6. You buy a toy for $6.98. You give the clerk a $10 bill. How much change should you receive?

   ○ **D.** $4.02
   ○ **E.** $3.02
   ○ **F.** $4.98

7. You are 53 inches tall. Your cousin is 46 inches tall. How much taller are you?

   ○ **A.** 13 inches
   ○ **B.** 99 inches
   ○ **C.** 7 inches

8. Your library book is 192 pages long. You have just finished page 75. How many pages do you have left to read?

   ○ **D.** 117
   ○ **E.** 123
   ○ **F.** 127

9. You have 185 days of school each year. Today is the 99th day of school. How many days of school do you have left for this year after today is over?

   ○ **A.** 114
   ○ **B.** 86
   ○ **C.** 84

PRACTICE ANSWERS
Page 82

1. 12
2. 61
3. 465
4. 541
5. 18
6. 19
7. 34
8. 26
9. 17
10. 37
11. 44
12. 531
13. 73
14. 496
15. 185
16. 283
17. 380
18. 131
19. 195
20. 192
21. 331
22. 353
23. 542
24. 841

Page 83

25.
$$\begin{array}{r} 594 \\ +283 \\ \hline 877 \end{array}$$

correct answer: 494

26.
$$\begin{array}{r} 361 \\ +435 \\ \hline 796 \end{array}$$

correct answer: 371

27.
$$\begin{array}{r} 216 \\ +680 \\ \hline 896 \end{array}$$

correct answer: 194

28.
$$\begin{array}{r} 246 \\ +196 \\ \hline 442 \end{array}$$

correct answer: 254

29. 469
30. 69
31. 255
32. 253
33. 18
34. 367
35. 10
36. 413
37. 337
38. 3963
39. 6339
40. 460
41. 936
42. 1630

Page 84

43. 0.2
44. 1.8
45. 0.7
46. 1.8
47. 0.6
48. 6.0 or 6
49. 2.35
50. 6.91
51. 3.27
52. 3.59
53. 1.52
54. 7.96
55. $3.44
56. $3.25
57. $5.70

TEST PREP ANSWERS
Page 85

1. B
2. D
3. C
4. D
5. B
6. E
7. C
8. D
9. B

Name _____ Date _____

Multiplication

2 × 143 = ▪

172–173

H	T	O	
1	4	3	
×		2	
		6	Multiply the ones.
	8	0	Multiply the tens.
2	0	0	Multiply the hundreds.
2	8	6	Add the partial products.

Use the grid to line up the digits by place value.
Find the product.

1. 2 × 43

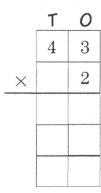

2. 2 × 213

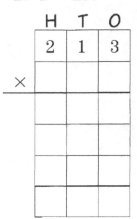

3. 4 × 102

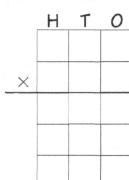

4. 3 × 233

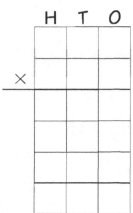

Multiply.

5.   21
    × 4

6.   14
    × 2

7.   143
    × 2

8.   223
    × 3

Name _____ Date _____

174-175 Use the grid to line up the digits by place value.
Find the product.

9. 3 × 24

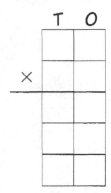

10. 6 × 82

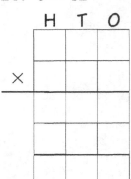

11. 5 × 37

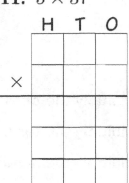

12. 4 × 39

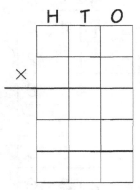

Multiply.

13. 36
      × 3

14. 25
      × 7

15. 41
      × 6

16. 79
      × 2

176-177

17. 614
      × 2

18. 265
      × 3

19. 826
      × 3

20. 421
      × 6

178

21. 203
      × 4

22. 621
      × 5

23. 480
      × 3

24. 250
      × 6

Name _____ Date _____

25. $1.59
     × 2

26. $6.27
     × 4

27. $3.05
     × 6

28. $0.98
     × 7

Use the prices to solve each problem.

Popcorn
Medium $1.75
Large $2.50
Jumbo $3.25

29. What will 3 medium popcorn boxes cost? _____

30. What will 2 large boxes of popcorn cost? _____

31. Will it cost more to buy 2 medium boxes of popcorn or 1 jumbo box?

_____

Use the grid to line up the digits by place value.
List all partial products and add to find the product.

32. 13 × 24

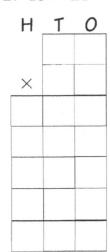

33. 25 × 37

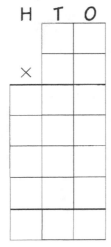

34. 34 × 46

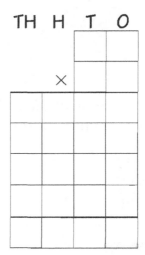

Name _____ Date _____

180–182 Use the grid to line up the digits by place value.
Find the product without listing every partial product.

35. 15 × 37

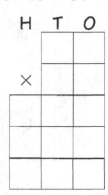

36. 24 × 46

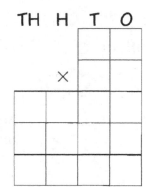

37. 42 × 38

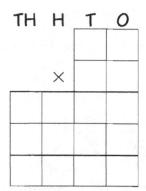

Multiply.

38.   24
        ×18

39.   43
        ×15

40.   21
        ×74

41.   23
        ×56

42.   43
        ×20

43.   22
        ×75

44.   38
        ×16

45.   56
        ×43

183 Check these products. If the answer is correct, circle correct. If the answer is incorrect, circle incorrect and write the correct product.

46.   42
          ×63
        _____
          126
          252
        _____
         2646

        correct
        incorrect: _____

47.   18
          ×53
        _____
           34
           50
        _____
          534

        correct
        incorrect: _____

48.   27
          ×34
        _____
          108
           81
        _____
          918

        correct
        incorrect: _____

Name _____   Date _____

Multiplication

Directions: Circle the letter that is beside the correct answer.

1. 413
    × 2

    A. 415
    B. 826
    C. 836
    D. 426

2. 23
    × 4

    F. 92
    G. 126
    H. 812
    J. 67

3. 258
    × 3

    A. 61,524
    B. 1214
    C. 764
    D. 774

4. 204
    × 5

    F. 1020
    G. 1000
    H. 2020
    J. 420

5. How many pictures can you take with 3 rolls of film?

    A. 62
    B. 72
    C. 612

6. How much will 4 rolls of film cost?

    F. $28.62
    G. $6.12
    H. $18.72

7. You want reprints of 14 different pictures. You want 6 reprints of each. How many reprints will you get altogether?

    A. 84
    B. 20
    C. 184

8. You mail pictures to relatives. You need to buy four 33¢ stamps. How much will the stamps cost?

    F. $1.32
    G. $12.12
    H. $1.32

Name _____ Date _____

Multiplication

Directions: Circle the letter beside the correct answer.
If the correct answer is not given, circle D.

1. 3×247
   A. 731
   B. 981
   C. 741
   D. None of these

2. 15×26
   A. 360
   B. 390
   C. 126
   D. None of these

3. 20×38
   A. 76
   B. 616
   C. 7600
   D. None of these

4. 31×32
   A. 992
   B. 118
   C. 690
   D. None of these

5. 12×74
   A. 192
   B. 160
   C. 888
   D. None of these

6. You are filling goodie bags for 16 children. You will put 5 wrapped candies in each bag. How many candies will you use?
   A. 200
   B. 530
   C. 80
   D. None of these

7. 12 chairs are in each row. There are 14 rows of chairs. How many chairs are there altogether?
   A. 168
   B. 60
   C. 142
   D. None of these

8. One package has 48 cookies. How many cookies will be in 15 packages?
   A. 880
   B. 720
   C. 180
   D. None of these

9. Cartons of ice cream cost $4.68 each. How much will 3 cartons cost?
   A. $18.44
   B. $34
   C. $14.04
   D. None of these

PRACTICE ANSWERS
Page 87
1. 86
2. 426
3. 408
4. 699
5. 84
6. 48
7. 286
8. 669

Page 88
9. 72
10. 492
11. 185
12. 156
13. 108
14. 175
15. 246
16. 158
17. 1228
18. 795
19. 2478
20. 2526
21. 812
22. 3105
23. 1440
24. 1500

Page 89
25. $3.18
26. $25.08
27. $18.30
28. $6.86
29. $5.25
30. $5
31. 2 medium boxes
32. 312
33. 925
34. 1564

Page 90
35. 555
36. 1104
37. 1596
38. 432
39. 645
40. 1554
41. 1288
42. 860
43. 1650
44. 608
45. 2408
46. correct
47. incorrect: 954
48. correct

TEST PREP A ANSWERS
Page 91
1. B
2. F
3. D
4. F
5. B
6. H
7. A
8. H

TEST PREP B ANSWERS
Page 92
1. C
2. B
3. D
4. A
5. C
6. C
7. A
8. B
9. C

Name _____ Date _____

Division

 Divide . If anything is left over, write it as the remainder.
Show your work.

1. $2\overline{)6}$ 2. $2\overline{)7}$ 3. $3\overline{)6}$ 4. $3\overline{)7}$

5. $5\overline{)15}$ 6. $5\overline{)16}$ 7. $4\overline{)10}$ 8. $6\overline{)14}$

9. $2\overline{)9}$ 10. $3\overline{)13}$ 11. $4\overline{)15}$ 12. $6\overline{)22}$

Solve.

13. You want to share 7 cookies evenly among 3 people.

    How many cookies will each person get? _____

    How many cookies are left over? _____

14. You have 13 muffins. You need to put 2 muffins in each basket.

    How many baskets do you need? _____

    How many muffins are left over? _____

15. 9 children need a ride to the movies. The cars can
    each take 4 children.

    How many cars do you need? _____
    (Make sure you have enough cars for all 9 children.
    No one can be left behind!)

Name _____ Date _____

186–187

H	T	O
2	1	3

```
    H  T  O
   ┌─────────
   │ 2  1  3
 3 │ 6  3  9
 - │ 6  0  0
   │    3  9
   │ -  3  0
   │       9
   │    -  9
   │       0
```

6 hundreds ÷ 3 = 2 hundreds
   3 × 2 hundreds = 600
   639 − 600 = 39

3 tens ÷ 3 = 1 ten
   3 × 1 ten = 30
   39 − 30 = 9

9 ones ÷ 3 = 3 ones
   3 × 3 ones = 9
   9 − 9 = 0

Divide. Show your work in the grid.

16.

H	T	O
2) 8	4	6
−		
	−	
		−

17.

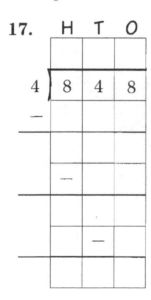

H	T	O
4) 8	4	8
−		
	−	
		−

18.

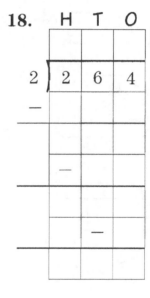

H	T	O
2) 2	6	4
−		
	−	
		−

Divide. Show your work.

19. 3)966   **20.** 2)482   **21.** 4)448   **22.** 3)939

Name _____ Date _____

188–189

```
      T   O
    ┌───┬───┐
    │ 2 │ 8 │   R 1
  ┌─┼───┼───┤
3 │ 8 │ 5 │   8 tens ÷ 3 = 2 tens and 2 tens left over
  └─┤   │   │       3 × 2 tens = 60
  ─ │ 6 │ 0 │       85 − 60 = 25
    ├───┼───┤
    │ 2 │ 5 │   25 ÷ 3 = 8 ones and 1 left over
  ─ │ 2 │ 4 │       3 × 8 ones = 24
    ├───┼───┤
    │   │ 1 │       25 − 24 = 1
    └───┴───┘
```

Divide. Show your work in the grids.

23. H T O

4) 9 4 3

24. H T O

3) 7 5 4

25. H T O

3) 6 4 6

Divide. Show your work.

26. 4)75   **27.** 3)637   **28.** 2)465   **29.** 3)589

Name _____ Date _____

Use an X to mark the place where you put the first digit in the quotient.

30. 8)648   **31.** 6)360   **32.** 9)546   **33.** 3)849

Divide. Show your work. You can use these facts.

7 ×0 / 0	7 ×1 / 7	7 ×2 / 14	7 ×3 / 21	7 ×4 / 28	7 ×5 / 35	7 ×6 / 42	7 ×7 / 49	7 ×8 / 56	7 ×9 / 63

34.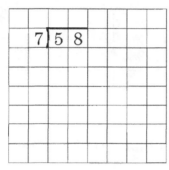

35.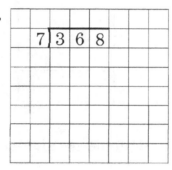

36.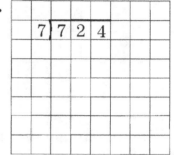

37.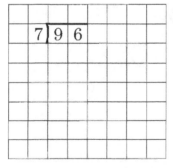

38.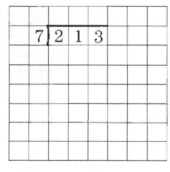

39.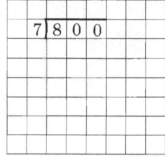

40.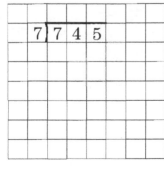

41.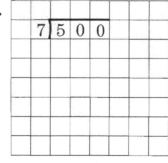

42.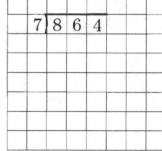

Name _____ Date _____

192-193 Divide the money. Show your work in the grids.

| $D = Dollars | d = dimes | p = pennies |

43.

$D	.d	p
$	.	
$7	.8	4

(dividend divided by 4)

44.

$D	.d	p
$	.	
$5	.1	9

(dividend divided by 3)

45.

$D	.d	p
$	.	
$9	.4	8

(dividend divided by 2)

Divide. Remember to show the dollar sign and decimal point.

46. 8)$4.72   **47.** 4)$8.64   **48.** 3)$0.24   **49.** 5)$7.35

50. 5)$1.85   **51.** 2)$3.14   **52.** 4)$4.40   **53.** 3)$6.99

Name _____ Date _____

Solve.   194–197

54. Three tennis balls come in a can.
17 children each need one tennis ball.
How many cans do you need to buy? _____

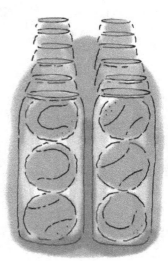

55. You have 260 marbles. You put
8 marbles in each bag. How many
marbles will you have left over? _____

56. You have 29 fluid ounces of fruit juice.
You pour it into 8-ounce glasses until
the pitcher is empty. How many glasses
do you fill?

_____

57. You pay $5 for 4 pounds of oranges.
How much does one pound of oranges cost? _____

Divide. Show your work. You can use these facts.   198–199

60 × 0 0	60 × 1 60	60 × 2 120	60 × 3 180	60 × 4 240	60 × 5 300	60 × 6 360	60 × 7 420	60 × 8 480	60 × 9 540

58. $60\overline{)775}$ **59.** $60\overline{)964}$ **60.** $60\overline{)458}$ **61.** $60\overline{)275}$

Name _____ Date _____

 Divide. Show your work.

62. $30\overline{)146}$ **63.** $70\overline{)234}$ **64.** $50\overline{)361}$ **65.** $20\overline{)289}$

 Divide. Show your work.

66. $5\overline{)425}$ **67.** $3\overline{)68}$ **68.** $6\overline{)253}$ **69.** $4\overline{)621}$

205 Divide. Then use multiplication or multiplication and addition to check your answer.

70. $5\overline{)225}$ **71.** $4\overline{)51}$

 Use the divisibility table in *Math to Know*, page 207 to help you answer these questions.

72. Is 135 divisible by 5? _____

How do you know? _____

73. Is 222 divisible by

2? _____   3? _____   4? _____   5? _____   6? _____   9? _____   10? _____

Name _____   Date _____

Division

Directions: Fill in the circle next to the letter of the correct answer.

1. $4\overline{)23}$

  ○ **A.** 5 R3

  ○ **B.** 4 R3

  ○ **C.** 2 R3

2. $2\overline{)628}$

  ○ **F.** 328

  ○ **G.** 304

  ○ **H.** 314

3. $3\overline{)76}$

  ○ **A.** 18 R2

  ○ **B.** 18

  ○ **C.** 25 R1

4. $3\overline{)624}$

  ○ **F.** 28

  ○ **G.** 211 R1

  ○ **H.** 208

5. $3\overline{)\$1.29}$

  ○ **A.** $0.43

  ○ **B.** $0.36

  ○ **C.** $1.43

6. Four children are sharing 34 game coupons evenly. What is the greatest number of coupons each child can get?

  ○ **F.** 7

  ○ **G.** 8

  ○ **H.** 9

7. Game markers come in packs of 4. There are 27 children in your class. How many packs do you need so that everyone gets one marker?

  ○ **A.** 7

  ○ **B.** 5

  ○ **C.** 6

8. You are dealing out 52 cards evenly to 5 children including yourself. How many cards will you have left over?

  ○ **F.** 1

  ○ **G.** 2

  ○ **H.** 3

9. You pay $9 for 4 pounds of cooked chicken wings. How much does 1 pound cost?

  ○ **A.** $2.00

  ○ **B.** $2.25

  ○ **C.** $2.50

Name _____ Date _____

Division

Directions: Solve each problem. Show your work.

1. $4\overline{)27}$ **2.** $3\overline{)639}$ **3.** $2\overline{)618}$ **4.** $3\overline{)\$1.26}$

5. Four children are sharing 26 game cards evenly. What is the greatest number of cards each child can get?

_____

6. Game markers come in packs of 4. There are 35 children in your class. How many packs do you need to buy so that everyone gets one marker?

_____

7. You are dealing out 26 cards evenly to 4 children including yourself. How many cards will you have left over?

_____

8. You pay $10 for 4 pounds of hamburger meat. How much does 1 pound cost?

_____

PRACTICE ANSWERS
Page 94
1. 3
2. 3 R1
3. 2
4. 2 R1
5. 3
6. 3 R1
7. 2 R2
8. 2 R2
9. 4 R1
10. 4 R1
11. 3 R3
12. 3 R4
13. 2 cookies, 1 cookie
14. 6 baskets, 1 muffin
15. 3 cars

Page 95
16. 423
17. 212
18. 132
19. 322
20. 241
21. 112
22. 313

Page 96
23. 235 R3
24. 251 R1
25. 215 R1
26. 18 R3
27. 212 R1
28. 232 R1
29. 196 R1

Page 97
30. $8\overline{)648}$ (X)
31. $6\overline{)360}$ (X)
32. $9\overline{)546}$ (X)
33. $3\overline{)849}$ (X)
34. 8 R2
35. 52 R4
36. 103 R3
37. 13 R5
38. 30 R3
39. 114 R2
40. 106 R3
41. 71 R3
42. 123 R3

Page 98
43. $1.96
44. $1.73
45. $4.74
46. $0.59
47. $2.16
48. $0.08
49. $1.47
50. $0.37
51. $1.57
52. $1.10
53. $2.33

Page 99
54. 6 cans
55. 4 marbles
56. $3\frac{5}{8}$ glasses
57. $1.25
58. 12 R55
59. 16 R4
60. 7 R38
61. 4 R35

Page 100
62. 4 R26
63. 3 R24
64. 7 R11
65. 14 R9
66. 85
67. 22 R2
68. 42 R1
69. 155 R1
70. 45 5 × 45 = 225
71. 12 R3 4 × 12 = 48
           48 + 3 = 51
72. Yes. The ones digit is 5.
73. yes, yes, no, no, yes, no, no

TEST PREP A ANSWERS
Page 101
1. A
2. H
3. C
4. H
5. A
6. G
7. A
8. G
9. B

TEST PREP B ANSWERS
Page 102
Check students' work.
1. 6 R3
2. 213
3. 309
4. $0.42
5. 4 cards
6. 9 packs
7. 1 card
8. $2.25

Tasty Yogurts

OBJECTIVE
- Compute with money and decimals

MATERIALS
- none

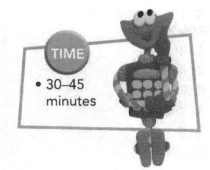

TIME
- 30–45 minutes

TEACHER NOTES
- Survey students to see how many eat yogurt.

- Discuss the prices and number of ounces on the pictures of yogurt.
  "Which carton costs the most? The least?"
  "Which has the greatest number of ounces? Least number of ounces?"

EXTENSIONS
- Make up more problems or have students make up problems using the data on their worksheet or the data in the table below.

What's in a 4-ounce serving?

Yogurt Name	Calories	Fat	Sugar
Yo-yo yogurt	110	0.5 g	19 g
Sprinkle On	130	1.0 g	21 g
Frogurt	140	3.5 g	18 g

- Have students write problems about food products they find in their homes. Have them copy information from the packages or bring the products or empty packages to school.

ANSWERS
1. 21¢
2. $1.52
3. $1.11
4. $2.72
5. 8.2 ounces
6. Divide 47¢ by 4.
7. 1.9 oz
8. Answers will vary. Check students' answers and reasons to support their answers.
9. 150 pounds

Name _____ Date _____

Tasty Yogurts

1. How much more does a container of Sprinkle On cost than a container of Yo-yo yogurt? _____

2. What is the total cost for one container of each yogurt? _____

3. How much do 3 containers of Frogurt cost? _____

4. How much do 4 containers of Sprinkle On cost?

    _____

5. How many ounces of yogurt are in 2 containers of Sprinkle On? _____

6. How would you find out how much one ounce of Yo-yo yogurt costs? _____

7. How much more yogurt is in a container of Sprinkle On than in a container of Frogurt? _____

8. Which yogurt do you think is more expensive—Yo-yo Yogurt of Frogurt? Tell how you decided.

    _____

    _____

9. In 1995, the average person in the Netherlands ate over 30 pounds of yogurt. How much yogurt did the average family of five in the Netherlands eat in 1995?

    _____

Name _____ Date _____

Fraction Concepts

210-213

1. In the fraction $\frac{2}{3}$, the numerator is _____ and the denominator is _____.

2. a. How many equal parts are in the circle with $\frac{2}{3}$ shaded? _____

b. How many of the equal parts are shaded? _____

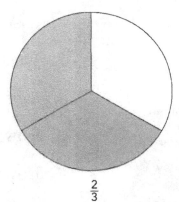

$\frac{2}{3}$

3. Shade $\frac{1}{4}$.   **4.** Shade $\frac{5}{6}$.   **5.** Shade $\frac{3}{3}$.

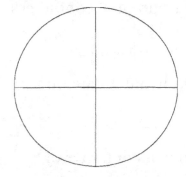 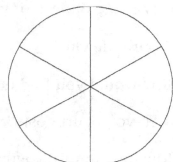 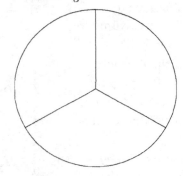

Write a fraction for the shaded part.

6. _____   **7.** _____   **8.** _____

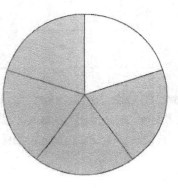 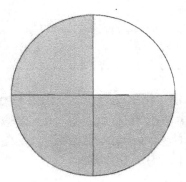 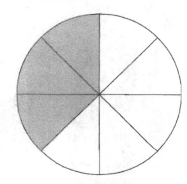

Name _____ Date _____

9. Label the number line with $\frac{1}{4}$, $\frac{2}{4}$, and $\frac{3}{4}$.

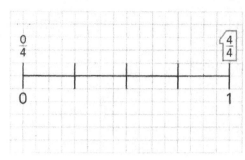

10. Circle $\frac{3}{5}$ of 5 and answer the question.

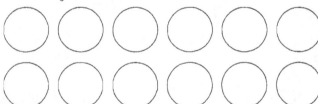

What is $\frac{3}{5}$ of 5? _____

11. Circle $\frac{1}{6}$ of 12.

What is $\frac{1}{6}$ of 12? _____

12. Circle $\frac{1}{3}$ of 6.

What is $\frac{1}{3}$ of 6? _____

Find the fraction of the group.

13. $\frac{1}{6}$ of 12 = _____   **14.** $\frac{2}{6}$ of 12 = _____   **15.** $\frac{5}{6}$ of 12 = _____

16. $\frac{3}{4}$ of 16 = _____   **17.** $\frac{1}{8}$ of 16 = _____   **18.** $\frac{2}{5}$ of 10 = _____

Math to Know

Name _____ Date _____

216   Tell how many are in one whole.

19. halves _____   **20.** thirds _____

21. fourths _____   **22.** sixths _____

217–219

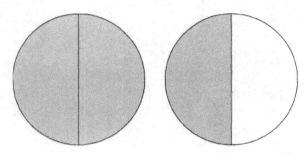

23. How many halves are in $1\frac{1}{2}$? _____

24. Write $1\frac{1}{2}$ as a fraction. _____

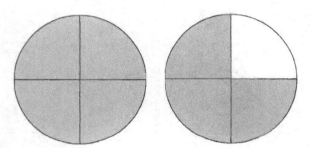

25. How many whole circles are in $\frac{11}{4}$? _____

26. Write $\frac{11}{4}$ as a mixed number. _____

Write the fraction as a whole or mixed number.

27. $\frac{4}{2}$ _____   **28.** $\frac{5}{2}$ _____   **29.** $\frac{4}{4}$ _____   **30.** $\frac{5}{4}$ _____   **31.** $\frac{7}{4}$ _____

Write the mixed number as a fraction.

32. $1\frac{1}{2}$ _____   **33.** $1\frac{1}{3}$ _____   **34.** $1\frac{2}{3}$ _____   **35.** $2\frac{1}{2}$ _____   **36.** $2\frac{3}{4}$ _____

Name _____ Date _____

37. Draw lines to connect equivalent fractions.

$\frac{1}{2}$　　　$\frac{3}{4}$　　　$\frac{2}{3}$　　　$\frac{1}{4}$　　　$\frac{1}{3}$

$\frac{4}{6}$　　　$\frac{2}{8}$　　　$\frac{2}{4}$　　　$\frac{3}{9}$　　　$\frac{6}{8}$

$\frac{3}{4} \times \boxed{\frac{2}{2}} = \frac{6}{8}$

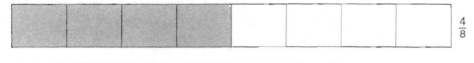

$\frac{3}{4}$

$\frac{6}{8}$

Complete to show how to write an equivalent fraction for each.

38. $\frac{3}{4} \times \boxed{\frac{\ }{\ }} = \frac{\square}{8}$　　**39.** $\frac{2}{3} \times \boxed{\frac{\ }{\ }} = \frac{\square}{12}$　　**40.** $\frac{3}{8} \times \boxed{\frac{\ }{\ }} = \frac{\square}{16}$

$\frac{4}{8} \div \frac{4}{4} = \frac{1}{2}$

$\frac{4}{8}$

$\frac{1}{2}$

Complete to show how to write the fraction in simplest form.

41. $\frac{4}{8} \div \boxed{\frac{\ }{\ }} = \frac{\square}{\square}$　　**42.** $\frac{2}{6} \div \boxed{\frac{\ }{\ }} = \frac{\square}{\square}$　　**43.** $\frac{6}{8} \div \boxed{\frac{\ }{\ }} = \frac{\square}{\square}$

Name _____ Date _____

224–225 Compare. Write < or >.

44. $\frac{3}{6}$ ◯ $\frac{5}{6}$

| $\frac{1}{6}$ | $\frac{1}{6}$ | $\frac{1}{6}$ | | | | $\frac{3}{6}$ |

| $\frac{1}{6}$ | $\frac{1}{6}$ | $\frac{1}{6}$ | $\frac{1}{6}$ | $\frac{1}{6}$ | | $\frac{5}{6}$ |

45. $\frac{3}{4}$ ◯ $\frac{2}{4}$

| $\frac{1}{4}$ | $\frac{1}{4}$ | $\frac{1}{4}$ | | $\frac{3}{4}$ |

| $\frac{1}{4}$ | $\frac{1}{4}$ | | | $\frac{2}{4}$ |

46. $\frac{5}{8}$ ◯ $\frac{7}{8}$ 47. $\frac{2}{3}$ ◯ $\frac{1}{3}$ 48. $\frac{5}{12}$ ◯ $\frac{7}{12}$

49. $\frac{1}{3}$ ◯ $\frac{1}{4}$

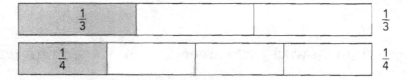

50. $\frac{1}{3}$ ◯ $\frac{1}{2}$

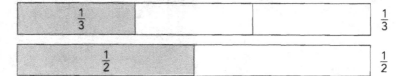

51. $\frac{1}{2}$ ◯ $\frac{1}{6}$ 52. $\frac{1}{6}$ ◯ $\frac{1}{4}$ 53. $\frac{1}{8}$ ◯ $\frac{1}{3}$

54. $\frac{2}{3}$ ◯ $\frac{4}{8}$ 55. $\frac{3}{4}$ ◯ $\frac{7}{8}$ 56. $\frac{5}{6}$ ◯ $\frac{7}{12}$

Order the fractions from smallest to largest.

57. $\frac{3}{4}, \frac{5}{12}, \frac{1}{2}$ 58. $\frac{1}{4}, \frac{1}{3}, \frac{1}{6}$

_____   _____

Name _____ Date _____

Fraction Concepts

Directions: Circle the letter of the correct answer.

1. What fraction of the circle is shaded?

A. $\frac{1}{8}$

B. $\frac{3}{8}$

C. $\frac{5}{8}$

D. $\frac{8}{8}$

2. Which point is marked on the number line?

F. $\frac{1}{4}$

G. $\frac{2}{4}$

H. $\frac{3}{4}$

J. 3

3. What is $\frac{1}{2}$ of 6?

A. 1

B. 2

C. $\frac{1}{3}$

D. 3

4. Which fraction is equal to 1?

F. $\frac{4}{4}$

G. $\frac{1}{4}$

H. $\frac{4}{1}$

J. $\frac{1}{2}$

5. Which mixed number describes the shaded parts of the circles?

A. $7\frac{1}{3}$

B. $3\frac{1}{3}$

C. $2\frac{1}{3}$

D. $2\frac{2}{3}$

6. Which fraction is equal to $\frac{1}{2}$?

F. $\frac{2}{3}$

G. $\frac{2}{4}$

H. $\frac{2}{5}$

J. 4

7. Which fraction shows the simplest form for $\frac{4}{12}$?

A. $\frac{1}{4}$

B. 12

C. $\frac{1}{2}$

D. $\frac{1}{3}$

8. Which fraction is greater than $\frac{1}{2}$?

F. $\frac{1}{3}$

G. $\frac{2}{3}$

H. $\frac{1}{4}$

J. $\frac{2}{4}$

Name _____ Date _____

Fraction Concepts

Directions: Write the letter of the correct answer.

1. What part of the figure is shaded?

A. $\frac{1}{8}$

B. $\frac{3}{8}$

C. $\frac{5}{8}$

D. $\frac{7}{8}$

2. Which point is marked on the number line?

F. $\frac{1}{3}$

G. $\frac{2}{3}$

H. $\frac{3}{2}$

J. 3

3. What is $\frac{1}{4}$ of 8?

A. 1

B. 2

C. 4

D. 6

4. Which fraction is equal to 1?

F. $\frac{5}{5}$

G. $\frac{1}{5}$

H. $\frac{5}{1}$

J. $\frac{1}{3}$

5. Which mixed number describes the shaded parts of the circles?

A. $\frac{5}{4}$

B. $3\frac{3}{4}$

C. $2\frac{3}{4}$

D. $2\frac{2}{4}$

6. Which fraction is equivalent to $\frac{1}{2}$?

F. $\frac{2}{3}$

G. $\frac{3}{6}$

H. $\frac{2}{6}$

J. 4

7. Which fraction shows the simplest form for $\frac{3}{12}$?

A. $\frac{1}{4}$

B. 12

C. $\frac{1}{2}$

D. $\frac{1}{3}$

8. Which fraction is greater than $\frac{1}{2}$?

F. $\frac{1}{3}$

G. $\frac{3}{4}$

H. $\frac{1}{4}$

J. $\frac{2}{4}$

PRACTICE ANSWERS
Page 106
1. 2, 3
2. a. 3
   b. 2
3.
4.
5.

6. $\frac{4}{5}$
7. $\frac{3}{4}$
8. $\frac{3}{8}$

Page 107
9.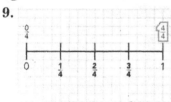

10. 3 of the 5 triangles should be circled; 3
11. 2 of the 12 circles should be circled; 2
12. 2 of the 6 squares should be circled; 2
13. 2
14. 4
15. 10
16. 12
17. 2
18. 4

Page 108
19. 2
20. 3
21. 4
22. 6
23. 3
24. $\frac{3}{2}$
25. 2
26. $2\frac{3}{4}$
27. 2
28. $2\frac{1}{2}$
29. 1
30. $1\frac{1}{4}$
31. $1\frac{3}{4}$
32. $\frac{3}{2}$
33. $\frac{4}{3}$
34. $\frac{5}{3}$
35. $\frac{5}{2}$
36. $\frac{11}{4}$

Page 109
37.

38. $\frac{3}{4} \times \frac{2}{2} = \frac{6}{8}$
39. $\frac{2}{3} \times \frac{4}{4} = \frac{8}{12}$
40. $\frac{3}{8} \times \frac{2}{2} = \frac{6}{16}$
41. $\frac{4}{8} \div \frac{4}{4} = \frac{1}{2}$
42. $\frac{2}{6} \div \frac{2}{2} = \frac{1}{3}$
43. $\frac{6}{8} \div \frac{2}{2} = \frac{3}{4}$

Page 110
44. <
45. >
46. <
47. >
48. <
49. >
50. <
51. >
52. <
53. <
54. >
55. <
56. >
57. $\frac{5}{12}, \frac{1}{2}, \frac{3}{4}$
58. $\frac{1}{6}, \frac{1}{4}, \frac{1}{3}$

TEST PREP A ANSWERS
Page 111
1. C
2. H
3. D
4. F
5. C
6. G
7. D
8. G

TEST PREP B ANSWERS
Page 112
1. C
2. G
3. B
4. F
5. C
6. G
7. A
8. G

Name _____ Date _____

Computing with Fractions

 227–228 Complete the addition sentence so it has a sum equivalent to $\frac{1}{2}$.
Do not use 0 as a numerator.

1. $\dfrac{\square}{4} + \dfrac{\square}{4} = \dfrac{2}{4} = \dfrac{1}{2}$ **2.** $\dfrac{\square}{6} + \dfrac{\square}{6} = \dfrac{3}{6} = \dfrac{1}{2}$ **3.** $\dfrac{\square}{8} + \dfrac{\square}{8} = \dfrac{4}{8} = \dfrac{1}{2}$

Add. Write the sum in simplest form.

4. $\dfrac{1}{4}$ **5.** - $\dfrac{5}{8}$ **6.** $\dfrac{3}{12}$

    $+ \dfrac{2}{4}$ $+ \dfrac{1}{8}$ $+ \dfrac{4}{12}$

7. $\dfrac{2}{5}$ **8.** $\dfrac{2}{6}$ **9.** $\dfrac{5}{10}$

    $+ \dfrac{2}{5}$ $+ \dfrac{3}{6}$ $+ \dfrac{2}{10}$

10. $\dfrac{1}{6}$ **11.** $\dfrac{2}{8}$ **12.** $\dfrac{3}{16}$

    $+ \dfrac{2}{6}$ $+ \dfrac{3}{8}$ $+ \dfrac{5}{16}$

229 Complete the addition sentences so they have sums equivalent to $\frac{1}{2}$.
Do not use 0 as a numerator. Color the models to help you.

13.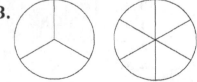

$\dfrac{\square}{3} \quad + \quad \dfrac{\square}{6} \quad = \quad \dfrac{1}{2}$

14.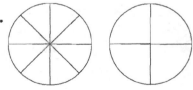

$\dfrac{\square}{8} \quad + \quad \dfrac{\square}{4} \quad = \quad \dfrac{1}{2}$

Name _____ Date _____

Complete the addition sentence so it has a sum equivalent to $\frac{1}{2}$.
Do not use 0 as a numerator. Rewrite the addition sentence
with common denominators.

229

15.

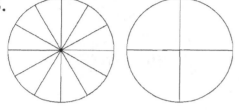

$$\frac{\square}{12} + \frac{\square}{4} = \frac{1}{2}$$

$$\frac{\square}{12} + \frac{\square}{12} = \frac{\square}{12}$$

16.

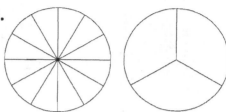

$$\frac{\square}{12} + \frac{\square}{3} = \frac{1}{2}$$

$$\frac{\square}{12} + \frac{\square}{12} = \frac{\square}{12}$$

Complete to find the sum.

Rewrite an addition sentence using equivalent fractions.

17. $\frac{1}{4} + \frac{1}{8} =$

$$\frac{\square}{8} + \frac{\square}{8} = \frac{\square}{8}$$

18. $\frac{1}{4} + \frac{2}{3} =$

$$\frac{\square}{12} + \frac{\square}{12} = \frac{\square}{12}$$

19. $\frac{1}{2} + \frac{1}{6} =$

$$\frac{\square}{6} + \frac{\square}{6} = \frac{\square}{6} = \frac{\square}{\square}$$

20. $\frac{1}{6} + \frac{1}{3} =$

$$\frac{\square}{6} + \frac{\square}{6} = \frac{\square}{6} = \frac{\square}{\square}$$

Name _____ Date _____

229 Rewrite with like denominators. Add. Simplify.

21. $\frac{1}{2} + \frac{1}{4}$ **22.** $\frac{2}{3} + \frac{1}{6}$ **23.** $\frac{2}{5} + \frac{3}{10}$

24. $\frac{1}{3} + \frac{2}{9}$ **25.** $\frac{3}{8} + \frac{1}{4}$ **26.** $\frac{1}{6} + \frac{1}{2}$

27. $\frac{1}{2} + \frac{1}{10}$ **28.** $\frac{1}{3} + \frac{1}{6}$ **29.** $\frac{3}{8} + \frac{1}{4}$

30. $\frac{2}{6} + \frac{1}{2}$ **31.** $\frac{1}{2} + \frac{3}{8}$ **32.** $\frac{1}{12} + \frac{3}{4}$

230-231 Add. Write the sum in simplest form.

33. $1\frac{1}{4} + 2\frac{1}{4}$ **34.** $2\frac{3}{8} + 3\frac{2}{8}$ **35.** $1\frac{1}{6} + 4\frac{3}{6}$

36. $2\frac{1}{3} + 2\frac{3}{6}$ **37.** $1\frac{4}{6} + 4\frac{1}{12}$ **38.** $3\frac{1}{3} + 2\frac{1}{4}$

Name _____ Date _____

Subtract. Write the difference in simplest form.

`232`

39. $\dfrac{3}{4}$ **40.** $\dfrac{5}{8}$ **41.** $\dfrac{7}{12}$ **42.** $\dfrac{4}{5}$

$\quad -\dfrac{1}{4}$ $-\dfrac{1}{8}$ $-\dfrac{4}{12}$ $-\dfrac{1}{5}$

Rewrite the subtraction sentence using equivalent fractions. Find the difference.

`233`

43. $\dfrac{3}{4} \quad - \quad \dfrac{1}{8}$ **44.** $\dfrac{1}{2} \quad - \quad \dfrac{1}{6}$

$$\dfrac{\square}{8} + \dfrac{\square}{8} = \dfrac{\square}{8} \qquad\qquad \dfrac{\square}{6} - \dfrac{\square}{6} = \dfrac{\square}{6} = \dfrac{\square}{\square}$$

Subtract. Write the difference in simplest form.

45. $\dfrac{1}{2} - \dfrac{1}{4}$ **46.** $\dfrac{5}{8} - \dfrac{1}{4}$ **47.** $\dfrac{9}{10} - \dfrac{2}{5}$

48. $\dfrac{3}{4} - \dfrac{2}{8}$ **49.** $\dfrac{1}{3} - \dfrac{1}{6}$ **50.** $\dfrac{1}{3} - \dfrac{1}{4}$

51. $3\dfrac{3}{4} - 2\dfrac{1}{4}$ **52.** $2\dfrac{3}{8} - 1\dfrac{2}{8}$ **53.** $5\dfrac{5}{6} - 4\dfrac{3}{6}$ `234`

54. $2\dfrac{2}{3} - 1\dfrac{1}{6}$ **55.** $5\dfrac{4}{6} - 2\dfrac{5}{12}$ **56.** $3\dfrac{2}{3} - 1\dfrac{1}{4}$

Name _____ Date _____

Computing with Fractions

Directions: Circle the letter of the correct answer.

1. You mix $\frac{1}{3}$ cup lemonade and $\frac{1}{3}$ cup tea. How full is the cup?

   A. $\frac{2}{3}$ cup

   B. 2 cups

   C. 3 cups

2. You live $\frac{1}{4}$ mile from school. How far is your walk to school and back home?

   F. $\frac{1}{4}$ mile

   G. 1 mile

   H. $\frac{1}{2}$ mile

3. You ate $\frac{1}{4}$ of the pizza and your brother ate $\frac{1}{2}$ of the pizza. How much of the pizza is eaten altogether?

   A. $\frac{2}{4}$

   B. $\frac{3}{4}$

   C. $\frac{2}{6}$

4. $2\frac{1}{4}$ cups of lemonade added to $1\frac{1}{2}$ cups of grape juice makes how much juice?

   F. $3\frac{3}{4}$ cups

   G. $\frac{5}{6}$ cups

   H. $3\frac{2}{8}$ cups

5. You have $\frac{7}{8}$ cup of milk and $\frac{5}{8}$ cup of juice. How much more milk do you have?

   A. $1\frac{1}{2}$ cups

   B. 2 cups

   C. $\frac{2}{8}$ cup

6. At soccer practice, you ran for $\frac{1}{4}$ hour and scrimmaged for $\frac{1}{2}$ hour. How much longer did you scrimmage?

   F. $\frac{3}{4}$ hour

   G. $\frac{2}{6}$ hour

   H. $\frac{1}{4}$ hour

7. You are $4\frac{1}{2}$ feet tall. Ben is $3\frac{1}{4}$ feet tall. How much taller are you than Ben?

   A. $1\frac{1}{4}$ feet

   B. $1\frac{1}{2}$ feet

   C. 1 foot

8. You walked $\frac{1}{3}$ mile to your friend's house and $\frac{1}{4}$ mile to the park. How far did you walk altogether?

   F. $\frac{1}{12}$ mile

   G. $\frac{2}{7}$ mile

   H. $\frac{7}{12}$ mile

PRACTICE ANSWERS
Page 114

1. 1, 1
2. 1, 2 or 2, 1
3. 1, 3 or 3, 1 or 2, 2
4. $\frac{3}{4}$
5. $\frac{3}{4}$
6. $\frac{7}{12}$
7. $\frac{4}{5}$
8. $\frac{5}{6}$
9. $\frac{7}{10}$
10. $\frac{1}{2}$
11. $\frac{5}{8}$
12. $\frac{1}{2}$
13. 1, 1
14. 2, 1

Page 115

15. $\frac{3}{12} + \frac{1}{4} = \frac{1}{2}$
$$\frac{3}{12} + \frac{3}{12} = \frac{6}{12}$$

16. $\frac{2}{12} + \frac{1}{3} = \frac{1}{2}$
$$\frac{2}{12} + \frac{4}{12} = \frac{6}{12}$$

17. $\frac{1}{4} + \frac{1}{8} =$
$$\frac{2}{8} + \frac{1}{8} = \frac{3}{8}$$

18. $\frac{1}{4} + \frac{2}{3} =$
$$\frac{3}{12} + \frac{8}{12} = \frac{11}{12}$$

19. $\frac{1}{2} + \frac{1}{6} =$
$$\frac{3}{6} + \frac{1}{6} = \frac{4}{6} = \frac{2}{3}$$

20. $\frac{1}{6} + \frac{1}{3} =$
$$\frac{1}{6} + \frac{2}{6} = \frac{3}{6} = \frac{1}{2}$$

Page 116

21. $\frac{3}{4}$
22. $\frac{5}{6}$
23. $\frac{7}{10}$
24. $\frac{5}{9}$
25. $\frac{5}{8}$
26. $\frac{2}{3}$
27. $\frac{3}{5}$
28. $\frac{1}{2}$
29. $\frac{5}{8}$
30. $\frac{5}{6}$
31. $\frac{7}{8}$
32. $\frac{5}{6}$
33. $3\frac{1}{2}$
34. $5\frac{5}{8}$
35. $5\frac{2}{3}$
36. $4\frac{5}{6}$
37. $5\frac{3}{4}$
38. $5\frac{7}{12}$

Page 117

39. $\frac{1}{2}$
40. $\frac{1}{2}$
41. $\frac{1}{4}$
42. $\frac{3}{5}$
43. $\frac{1}{2} - \frac{1}{6} =$
$$\frac{3}{6} - \frac{1}{6} = \frac{2}{6} = \frac{1}{3}$$

44. $\frac{3}{4} - \frac{1}{8} =$
$$\frac{6}{8} - \frac{1}{8} = \frac{5}{8}$$

45. $\frac{1}{4}$
46. $\frac{3}{8}$
47. $\frac{1}{2}$
48. $\frac{1}{2}$
49. $\frac{1}{6}$
50. $\frac{1}{12}$
51. $1\frac{1}{2}$
52. $1\frac{1}{8}$
53. $1\frac{1}{3}$
54. $1\frac{1}{2}$
55. $3\frac{1}{4}$
56. $2\frac{5}{12}$

TEST PREP ANSWERS
Page 118

1. A
2. H
3. B
4. F
5. C
6. H
7. A
8. H

Fraction Strips

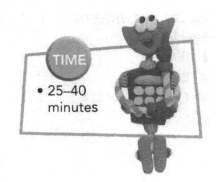

OBJECTIVES
- Make linear models for fractions
- Find equivalent fractions
- Compare fractions
- Add and subtract fractions

MATERIALS
For each student:
- 3 strips of paper ($4\frac{1}{4}$" x 11")

TIME
- 25–40 minutes

TEACHER NOTES
- Students will fold strips of paper into fourths, eighths, and sixths. They must fold a strip so that each section is the same size.

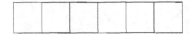

- By making parallel folds, the students will have linear fraction models. The fraction strips are very useful for finding equivalent fractions, comparing fractions, and computing with fractions.

EXTENSIONS
- Make fraction strips for halves, thirds, and twelfths.

- Use all six fraction strips for further work with finding equivalent fractions, comparing fractions, and computing with fractions.

- Make duplicate halves and fourths strips to work with mixed numbers. Ask questions such as:

How many halves are in a whole strip? (2)

$\frac{3}{2} = ?$ $(1\frac{1}{2})$

$\frac{4}{2} = ?$ (2)

How many fourths are in a whole strip? (4)

$\frac{5}{4} = ?$ $(1\frac{1}{4})$

$\frac{6}{4} = ?$ $(1\frac{2}{4} = 1\frac{1}{2})$

$\frac{3}{4} + \frac{3}{4} = ?$ $(\frac{6}{4} = 1\frac{2}{4} = 1\frac{1}{2})$

ANSWERS

1.

$\frac{1}{4}$	$\frac{1}{4}$	$\frac{1}{4}$	$\frac{1}{4}$

2.

$\frac{1}{8}$	$\frac{1}{8}$	$\frac{1}{8}$	$\frac{1}{8}$	$\frac{1}{8}$	$\frac{1}{8}$	$\frac{1}{8}$	$\frac{1}{8}$

3.

$\frac{1}{6}$	$\frac{1}{6}$	$\frac{1}{6}$	$\frac{1}{6}$	$\frac{1}{6}$	$\frac{1}{6}$

4. $\frac{1}{4}$

5. Rachel

6. 1 cup

7. Problems will vary.

Name _____ Date _____

Fraction Strips

Models for Fractions
pages 212-213

Fractions Equal to 1
page 216

Equivalent Fractions
page 220

Comparing Fractions
pages 224-225

Adding Fractions
pages 227-229

Subtracting Fractions
pages 232-233

1. Fold one of the strips of paper into fourths. Write $\frac{1}{4}$ in each section.

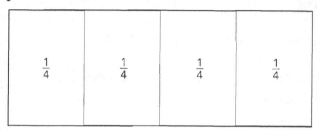

2. Fold a different strip of paper into eighths. Write $\frac{1}{8}$ in each section.

3. Fold another strip of paper into sixths. Write $\frac{1}{6}$ in each section.

Solve these problems. You may want to use your strips to help you.

4. Naomi ate $\frac{3}{4}$ of her sandwich. What fraction of her sandwich was left? _____

5. Ben and Rachel each bought a small pizza. Rachel ate $\frac{3}{4}$ of hers and Ben ate $\frac{4}{6}$ of his. Who ate more? _____

6. Andrew made a punch by mixing $\frac{3}{4}$ cup of lemonade with $\frac{2}{8}$ cup of orange juice. How much punch did he make? _____

7. Write a word problem that can be solved by using one or more of your strips. Solve your problem.

_____
_____
_____
_____

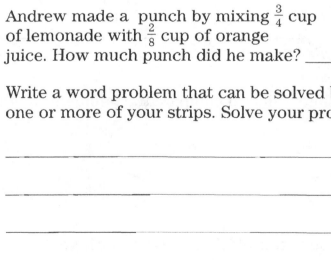

Name _____   Date _____

Positive and Negative Numbers

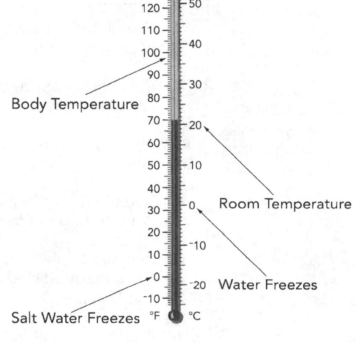

238–239   Write each using a signed number.

1. 10 degrees below zero Celsius _____

2. 20 degrees above zero Fahrenheit _____

The temperature goes down as it gets colder.
Tell which is colder.

3. −10°C or −20°C _____

4. −5°C or 10°C _____

Use the number line to compare the numbers. Write < or > in the ⃝.

5. 3 ◯ −1   **6.** −4 ◯ 2   **7.** 0 ◯ −1

8. −4 ◯ −2   **9.** −1 ◯ −5   **10.** −5 ◯ 5

11. Are you better off if you owe $5 or if you owe $3? _____

Name _____ Date _____

Positive and Negative Numbers

When you think about land elevations and ocean depths, you can think of sea level as zero. Land elevations are thought of as positive numbers above sea level. Ocean depths are thought of as negative numbers below sea level.

Mountain Elevations and Ocean Depths (Measurements are in feet.)			
Mountain	**Height**	**Ocean**	**Depth**
Mount McKinley	20,320	Pacific Ocean	−36,198
Mount Everest	29,028	Atlantic Ocean	−30,246
Mount Kilimanjaro	19,340	Indian Ocean	−24,460

Source: Time Almanac 1999

The chart lists the highest known peak of each mountain.

1. Which mountain peak is closest to sea level? _____

2. Which peak is farthest from sea level? _____

The chart lists the deepest trench in each ocean.

3. Which ocean trench depth shown is closest to sea level? _____

4. Which ocean has the trench farthest from sea level? _____

5. Complete the number line by filling in the blanks.

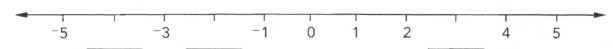

Compare. Write < or > in the ◯ .

6. 2 ◯ −2   7. −3 ◯ −4   8. −5 ◯ 3

9. Are you be better off if you owe $2 or if you owe $4? _____

PRACTICE ANSWERS
Page 122

1. $-10°C$
2. $20°C$ or $+20°C$
3. $-20°C$
4. $-5°C$
5. $>$
6. $<$
7. $>$
8. $<$
9. $>$
10. $<$
11. owe $3

TEST PREP ANSWERS
Page 123

1. Mount Kilimanjaro
2. Mount Everest
3. Indian Ocean
4. Pacific Ocean
5.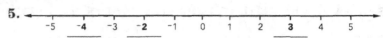
6. $>$
7. $>$
8. $<$
9. owe $2

Name _____ Date _____

Properties

Order Properties

240-241

When you change the order of addends you do not change the sum.

Find each sum.

1. $5 + 2 =$ _____　　　　**2.** $7 + 4 =$ _____　　　　**3.** $6 + 9 =$ _____

$2 + 5 =$ _____　　　　　　　　$4 + 7 =$ _____　　　　　　　　$9 + 6 =$ _____

4. $25 + 18 + 5 =$ _____　**5.** $21 + 37 + 9 =$ _____　**6.** $48 + 36 + 2 =$ _____

$25 + 5 + 18 =$ _____　　　　$21 + 9 + 37 =$ _____　　　　$48 + 2 + 36 =$ _____

When you change the order of factors you do not change the product.

Write the product.

7. $4 \times 3 =$ _____　　　　**8.** $7 \times 5 =$ _____　　　　**9.** $8 \times 6 =$ _____

$3 \times 4 =$ _____　　　　　　　$5 \times 7 =$ _____　　　　　　　$6 \times 8 =$ _____

10. $5 \times 36 \times 2 =$ _____　　　　　　**11.** $4 \times 13 \times 25 =$ _____

$5 \times 2 \times 36 =$ _____　　　　　　　　$4 \times 25 \times 13 =$ _____

Grouping Properties

242-243

When you change the grouping of three or more addends you do not change the sum.

Find the sum.

12. $(5 + 3) + 7 =$ _____　　　　　　**13.** $(37 + 5) + 45 =$ _____

$5 + (3 + 7) =$ _____　　　　　　　　$37 + (5 + 45) =$ _____

Name _____ Date _____

242-243 **When you change the grouping of three or more factors you do not change the product.**

Find the products.

14. $(7 \times 5) \times 2 =$ _____   **15.** $(3 \times 4) \times 5 =$ _____   **16.** $(2 \times 4) \times 10 =$ _____

     $7 \times (5 \times 2) =$ _____ $3 \times (4 \times 5) =$ _____ $2 \times (4 \times 10) =$ _____

244-245 **Distributive Property**
You can separate numbers into parts so that the numbers are easier to work with.

$3 \times 14 = \blacksquare$

               14 = 10 + 4

(3×10) + (3×4)
   30   +   12   = 42

17. $4 \times 23 = (4 \times 20) + (4 \times$ _____$) =$ _____ $+$ _____ $=$ _____

18. $5 \times 46 = (5 \times 40) + (5 \times$ _____$) =$ _____ $+$ _____ $=$ _____

19. $4 \times 29 = (4 \times 30) - (4 \times$ _____$) =$ _____ $+$ _____ $=$ _____

20. $7 \times 19 = (7 \times 20) - (7 \times$ _____$) =$ _____ $+$ _____ $=$ _____

246 **Adding 0 Property**
If you add 0 to any number, the sum is that same number.

Multiplying by 1 Property
If you multiply any number by 1, the product is that same number.

21. $5 + 0 =$ _____   **22.** $5 \times 1 =$ _____   **23.** $7 + 0 =$ _____

24. $26 + 0 =$ _____   **25.** $9 \times 1 =$ _____   **26.** $58 \times 1 =$ _____

Name _____ Date _____

Zero Property of Multiplication
If you multiply any number and 0, the product is 0.

27. $5 \times 0 =$ _____   **28.** $16 \times 0 =$ _____   **29.** $37 \times 6 \times 49 \times 0 =$ _____

The Addition Property of Equality
You can add the same number to both sides of an equation.

3 = 3

3 + 2 = 3 + 2

30. Add the same number to both sides of $3 = 3$.

$3 +$ ____ $= 3 +$ ____ $3 +$ ____ $= 3 +$ ____

The Multiplication Property of Equality
You can multiply both sides of an equation by the same number.

2 = 2

3 × 2 = 3 × 2

31. Multiply both sides of $2 = 2$ by the same number.

____ $\times 2 =$ ____ $\times 2$ ____ $\times 2 =$ ____ $\times 2$

Name _____ Date _____

Properties

Use the Order Property of Addition to rewrite. Then find the sum.

1. $8 + 4 + 2$ **2.** $65 + 18 + 5$

_____ = _____   _____ = _____

Use the Order Property of Multiplication to rewrite. Then find the product.

3. $5 \times 37 \times 2$ **4.** $4 \times 3 \times 5$

_____ = _____   _____ = _____

Use the Grouping Property of Addition to rewrite. Then find the sum.

5. $(6 + 7) + 3$ **6.** $(27 + 16) + 4$

_____ = _____   _____ = _____

Use the Grouping Property of Multiplication to rewrite.
Then find the product.

7. $(7 \times 5) \times 2$ **8.** $(25 \times 15) \times 4$

_____ = _____   _____ = _____

9. Use the Distributive Property to complete.

$4 \times 16 = (4 \times 10) + (4 \times \underline{\hspace{1cm}}) = \underline{\hspace{1cm}}$

10. What happens when you add zero to a number?

_____

11. What happens when you multiply a number by one?

_____

12. What happens when you multiply a number by zero?

_____

PRACTICE ANSWERS
Page 125

1. 7, 7
2. 11, 11
3. 15, 15
4. 48, 48
5. 67, 67
6. 86, 86
7. 12, 12
8. 35, 35
9. 48, 48
10. 360, 360
11. 1300, 1300
12. 15, 15
13. 87, 87

Page 126

14. 70, 70
15. 60, 60
16. 80, 80
17. $(4 \times 20) + (4 \times \underline{3}) = \underline{80} + \underline{12} = \underline{92}$
18. $(5 \times 40) + (5 \times \underline{6}) = \underline{200} + \underline{30} = \underline{230}$
19. $(4 \times 30) - (4 \times \underline{1}) = \underline{120} - \underline{4} = \underline{116}$
20. $(7 \times 20) - (7 \times \underline{1}) = \underline{140} - \underline{7} = \underline{133}$
21. 5
22. 5
23. 7
24. 26
25. 9
26. 58

Page 127

27. 0
28. 0
29. 0
30. Answers will vary.
Possible answers:

$3 + \underline{1} = 3 + \underline{1}$
$3 + \underline{2} = 3 + \underline{2}$
$3 + \underline{3} = 3 + \underline{3}$
$3 + \underline{4} = 3 + \underline{4}$

31. Answers will vary.
Possible answers:

$\underline{1} \times 2 = \underline{1} \times 2$
$\underline{2} \times 2 = \underline{2} \times 2$
$\underline{3} \times 2 = \underline{3} \times 2$
$\underline{4} \times 2 = \underline{4} \times 2$

TEST PREP ANSWERS
Page 128

Answers may vary.
Possible answers:

1. $8 + 2 + 4 = 14$
2. $65 + 5 + 18 = 88$
3. $5 \times 2 \times 37 = 370$
4. $4 \times 5 \times 3 = 60$
5. $6 + (7 + 3) = 16$
6. $27 + (16 + 4) = 47$
7. $7 \times (5 \times 2) = 70$
8. $(25 \times 4) \times 15 = 150$
9. $4 \times 16 = (4 \times 10) + (4 \times \underline{6}) = \underline{64}$
10. The sum is the number.
11. The product is the number.
12. The product is zero.

Name _____ Date _____

Expressions

250–252 Write an expression for each person's age in years.
Let c = Cindy's age in years.

1. Jay is 2 years older than Cindy. _____

2. Renee is 3 years younger than Cindy. _____

3. Sandy is three times as old as Cindy. _____

Write an expression for each person's height in inches.
Let s = Sean's height in inches.

4. Chris is 3 inches taller than Sean. _____

5. Mrs. Stebbins is twice as tall as Sean. _____

6. Will is the same height as Sean. _____

Write an expression for each girl's age in years.
Let d = Dana's age in years.

7. Jane is the same age as Dana. _____

8. Roberta is 6 years older than Dana. _____

9. Nikki is 5 times as old as Dana. _____

10. Emma is 4 years younger than Dana. _____

Write an expression for the number of marbles each girl has.
Let b = the number of marbles in a full bag.

11. Katy has one full bag of marbles. _____

12. Jana has one full bag and 3 extra marbles. _____

13. Emily has two full bags. _____

14. Lauren had a full bag until she lost 5 marbles. _____

Name _____ Date _____

Rules for Order of Operations:
1. Multiply and divide in the order they appear.
2. Add and subtract in the order they appear.

253

15. $40 \div 10 + 6 \times 2 =$ _____   **16.** $8 \times 2 - 4 + 3 =$ _____

17. $5 + 4 \div 2 - 1 =$ _____   **18.** $3 + 2 \times 5 - 1 =$ _____

19. $7 \div 7 + 7 - 3 =$ _____   **20.** $6 + 2 \times 3 + 3 =$ _____

21. $9 - 4 \times 0 + 5 =$ _____   **22.** $7 + 9 \times 1 \div 3 =$ _____

23. $8 - 2 + 4 - 8 \times 0 =$ _____   **24.** $19 - 6 \times 2 \div 4 =$ _____

25. $2 + 2 \times 8 + 9 =$ _____   **26.** $20 + 6 \times 3 \div 2 =$ _____

When you have parentheses or exponents, do those first.

254

27. $2 + (7 - 3) \times 5 =$ _____   **28.** $5 + 3^2 \times 2 =$ _____

29. $4 + 5 \times (2 + 1) - 4^2 =$ _____   **30.** $9 + 6 \times 2^3 - 13 =$ _____

31. $100 - (7 - 3) \times 10 =$ _____   **32.** $1 + 2^3 \times 3 =$ _____

33. $18 \times (2 - 1) - 4^2 + 2 =$ _____   **34.** $7 \times 2^2 + (1 + 11) =$ _____

Name _____ Date _____

Expressions

Directions: Circle the letter of the correct answer.

1. There are 4 more girls than boys in the class. If b = the number of boys in the class, which expression shows the number of girls in the class?

   A. $b + 4$ B. $4 \times b$

   C. $b - 4$ D. $b \div 4$

2. Maria is two years younger than Bill. If b = Bill's age in years, which expression shows Maria's age in years?

   E. $2 \times b$ F. $2 + b$

   G. $b - 2$ H. $b \div 2$

3. Bob is 4 times as old as his son Gregory. If b = Bob's age in years, which expression shows Gregory's age in years?

   A. $b + 4$ B. $b - 4$

   C. $4 \times b$ D. $b \div 4$

4. Let b = the number of students on the bus. Six students get off. Which expression shows the number of students left on the bus?

   E. $b \times 6$ F. $b - 6$

   G. $6 - b$ H. $6 \div b$

5. $4 + 3 \times 2 - 1$

   A. 7

   B. 9

   C. 13

   D. 14

6. $6 + (4 - 1) \times 2^2$

   E. 18

   F. 324

   G. 36

   H. 72

7. $14 - 3 + 6 \times 1$

   A. 17

   B. 9

   C. 5

   D. 24

8. $12 - 2^3 + (4 - 1)$

   E. 1003

   F. 7

   G. 1005

   H. 9

PRACTICE ANSWERS
Page 130
1. $c + 2$
2. $c - 3$
3. $3 \times c$ or $3c$
4. $s + 3$
5. $2 \times s$ or $2s$
6. s
7. d
8. $d + 6$
9. $5 \times d$ or $5d$
10. $d - 4$
11. b
12. $b + 3$
13. $2 \times b$ or $2b$
14. $b - 5$

Page 131
15. 16
16. 15
17. 6
18. 12
19. 5
20. 15
21. 14
22. 10
23. 10
24. 16
25. 27
26. 29
27. 22
28. 23
29. 3
30. 44
31. 60
32. 25
33. 4
34. 40

TEST PREP ANSWERS
Page 132
1. A
2. G
3. D
4. F
5. B
6. E
7. A
8. F

Name _____ Date _____

Equations

255–257

1. Look at this scale. It is balanced. Both sides have the same number of blocks.

   Let x = the number of blocks in the bag.

   a. From the scale, you can write the equation $x + 2 =$ _____.

   b. Solve the equation you wrote. $x =$ _____

   c. How many blocks are in the bag? _____

2. Now look at this scale. The same number of blocks are in each bag.

   Let y = the number of blocks in each bag.

   a. Use the scale to write an equation. _____.

   b. Solve the equation you wrote. $y =$ _____

   c. How many blocks are in each bag? _____

Solve the equation.

3. $x + 6 = 15$ **4.** $9 + y = 13$ **5.** $z - 5 = 7$

    $x =$ _____ $y =$ _____ $z =$ _____

6. $9 - t = 7$ **7.** $14 - m = 7$ **8.** $n - 7 = 5$

    $t =$ _____ $m =$ _____ $n =$ _____

9. $6 \times p = 24$ **10.** $8 \times m = 40$ **11.** $k + 8 = 73$

    $p =$ _____ $m =$ _____ $k =$ _____

12. $18 + q = 72$ **13.** $m + 57 = 60$ **14.** $t \times 25 = 125$

    $q =$ _____ $m =$ _____ $t =$ _____

Name _____   Date _____

15. Wendy paid $14 for a tape and a CD. The CD cost $9.
How much did the tape cost?

     a. Circle the equation you could use to solve the problem.

        $t + 9 = 14$ $14 + 9 = t$ $t - 9 = 14$

     b. Solve the equation. $t =$ _____

     c. How much did the tape cost? _____

16. Nancy bought 3 bottles of water for 75¢. Each cost the same.
How much did each bottle cost?

     a. Circle the equation you could use to solve the problem.

        $3 \times 75 = b$ $b - 3 = 75$ $3 \times b = 75$

     b. Solve the equation. $b =$ _____

     c. How much did each bottle cost? _____

17. Sandy is 4 years younger than Max. Sandy is 9 years old.
How old is Max?

     a. Circle the equation you could use to solve the problem.

        $m + 4 = 9$ $4 - m = 9$ $m - 4 = 9$

     b. Solve the equation. $m =$ _____

     c. How old is Max? _____

18. Jessica spends 14 hours each week practicing the piano. What is
the average number of hours she practices each day?

     a. Circle the equation you could use to solve the problem.

        $7 \times a = 14$ $a - 7 = 14$ $7 + a = 14$

     b. Solve the equation. $a =$ _____

     c. What is the average number of hours she practices each day? _____

Name _____   Date _____

Equations

Directions: Circle the letter next to the best answer.

1. Together a notebook and a pen cost $6. The pen cost $2. Which equation could you solve to find the cost of the notebook?

   A. $n - 6 = 2$

   B. $n + 2 = 6$

   C. $n - 2 = 6$

   D. $2 \times n = 6$

2. Samantha is 2 inches shorter than Ben. Samantha is 48 inches tall. Which equation could you solve to find Ben's height?

   A. $b - 2 = 48$

   B. $b + 2 = 48$

   C. $b - 48 = 2$

   D. $2 \times b = 48$

3. Three times a number is 12. Which equation could you solve to find the number?

   A. $3 + n = 12$

   B. $n - 3 = 12$

   C. $3 - n = 12$

   D. $3 \times n = 12$

4. Solve. $x + 7 = 13$

   A. $x = 7$

   B. $x = 6$

   C. $x = 20$

   D. $x = 13$

5. Solve. $16 - y = 9$

   A. $y = 7$

   B. $y = 25$

   C. $y = 9$

   D. $y = 8$

6. Solve. $m - 8 = 32$

   A. $m = 4$

   B. $m = 40$

   C. $m = 24$

   D. $m = 32$

7. Solve. $6 \times r = 48$

   A. $r = 48$

   B. $r = 8$

   C. $r = 42$

   D. $r = 288$

PRACTICE ANSWERS
Page 134

1. **a.** 6
   b. 4
   c. 4 blocks
2. **a.** $2 \times y = 6$ or $2y = 6$
   b. 3
   c. 3 blocks
3. 9
4. 4
5. 12
6. 2
7. 7
8. 12
9. 4
10. 5
11. 65
12. 54
13. 3
14. 5

Page 135

15. **a.** $t + 9 = 14$
    b. 5
    c. $5
16. **a.** $3 \times b = 75$
    b. 25
    c. 25¢
17. **a.** $m - 4 = 9$
    b. 13
    c. 13 years old
18. **a.** $7 \times a = 14$
    b. 2
    c. 2 hours

TEST PREP ANSWERS
Page 136

1. B
2. A
3. D
4. B
5. A
6. B
7. B

Name _____ Date _____

Graphing Ordered Pairs

258 An ordered pair of numbers is used to name a location on a grid. The first number tells you the horizontal position. The second number tells you the vertical position.

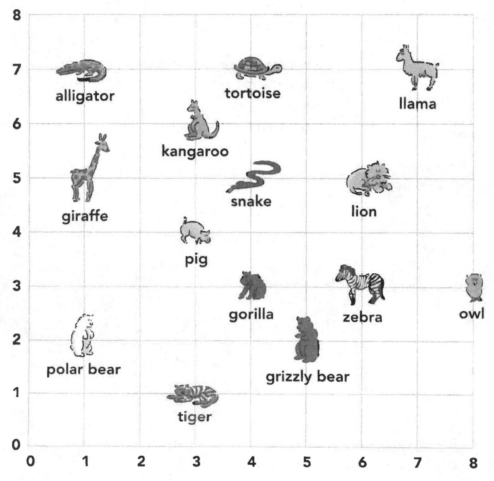

What animal is at each ordered pair?

1. (3, 1) _____

2. (4, 7) _____

3. (1, 5) _____

4. (6, 5) _____

5. (8, 3) _____

6. (5, 2) _____

7. (1, 2) _____

8. (4, 3) _____

9. (7, 7) _____

10. (6, 3) _____

11. (1, 7) _____

12. (3, 6) _____

13. (4, 5) _____

14. (3, 4) _____

Name _____ Date _____

You can name a specific location on a grid with an ordered pair of numbers. The first number describes the horizontal position and the second number describes the vertical position.

259

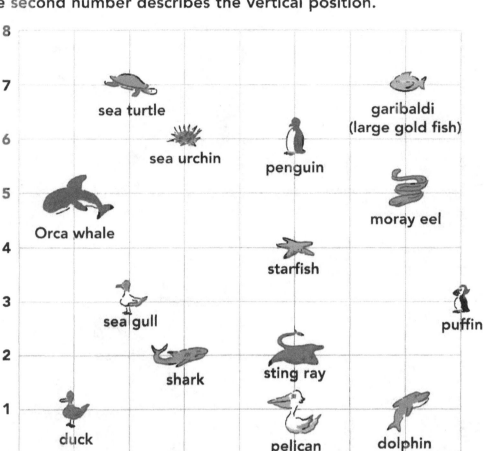

Write the ordered pair that tells the location of the water animal.

15. dolphin _____   16. shark _____   17. Orca whale _____

18. penguin _____   19. sea turtle _____   20. sting ray _____

21. moray eel _____   22. garibaldi _____   23. pelican _____

24. sea gull _____   25. puffin _____   26. sea urchin _____

27. starfish _____   28. duck _____

Name _____ Date _____

Graphing Ordered Pairs

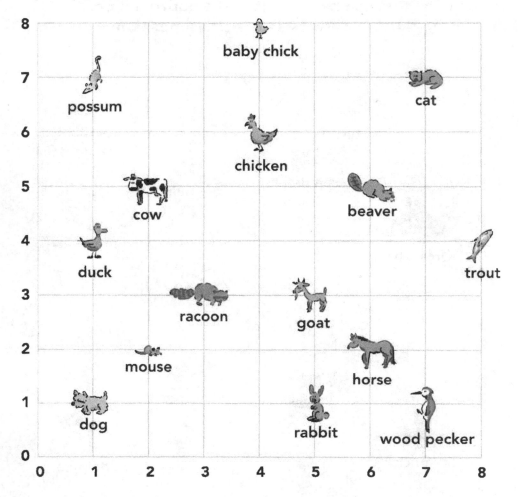

What picture is at the ordered pair?

   1. (5, 1) _____   **2.** (4, 6) _____   **3.** (6, 2) _____

   4. (5, 3) _____   **5.** (1, 4) _____   **6.** (4, 8) _____

Write the ordered pair that tells the location of each.

   7. beaver _____   **8.** trout _____   **9.** woodpecker _____

   10. dog _____   **11.** possum _____   **12.** cow _____

   13. raccoon _____   **14.** mouse _____   **15.** cat _____

PRACTICE ANSWERS
Page 138
1. tiger
2. tortoise
3. giraffe
4. lion
5. owl
6. grizzly bear
7. polar bear
8. gorilla
9. llama
10. zebra
11. alligator
12. kangaroo
13. snake
14. pig

Page 139
15. (7, 1)
16. (3, 2)
17. (1, 5)
18. (5, 6)
19. (2, 7)
20. (5, 2)
21. (7, 5)
22. (7, 7)
23. (5, 1)
24. (2, 3)
25. (8, 3)
26. (3, 6)
27. (5, 4)
28. (1, 1)

TEST PREP ANSWERS
Page 140
1. rabbit
2. chicken
3. horse
4. goat
5. duck
6. baby chick
7. (6, 5)
8. (8, 4)
9. (7, 1)
10. (1, 1)
11. (1, 7)
12. (2, 5)
13. (3, 3)
14. (2, 2)
15. (7, 7)

Name _____ Date _____

Functions

260–261 Fill in the blanks in the function table.

1.

Rule: + 2	
Input	Output
0	
1	
2	
	5
	6
5	
	8

2.

Rule: − 2	
Input	Output
2	
5	
9	
	4
	8
13	
	16

3.

Rule: × 2	
Input	Output
3	
5	
6	
	16
	18
13	
	32

Fill in the blanks in the function table and find the missing rule.

4.

Rule: _____	
Input	Output
1	4
4	7
6	9
7	
	12
12	
	19

5.

Rule: _____	
Input	Output
3	13
8	18
16	26
35	
	67
58	
	82

6.

Rule: _____	
Input	Output
2	10
4	20
5	25
7	
	45
11	
	60

Name _____ Date _____

Functions

Fill in the blanks in the function table.

1.

Rule: + 5	
Input	Output
0	
1	
2	
3	
	9
5	
	11

2.

Rule: − 1	
Input	Output
5	4
7	6
15	14
37	
	68
56	
	84

3.

Rule: × 5	
Input	Output
2	
3	
6	
	35
	45
12	
	75

Fill in the blanks in the function table and find the missing rule.

4.

Rule: _____	
Input	Output
3	5
5	7
6	8
7	
	14
15	
	18

5.

Rule: _____	
Input	Output
4	13
7	16
23	32
5	
	17
62	
	83

6.

Rule: _____	
Input	Output
3	14
5	16
26	37
28	
	54
67	
	89

PRACTICE ANSWERS
Page 142

1.

Rule: + 2	
Input	Output
0	**2**
1	**3**
2	**4**
3	5
4	6
5	**7**
6	8

2.

Rule: − 2	
Input	Output
2	**0**
5	**3**
9	**7**
2	4
6	8
13	**11**
14	16

3.

Rule: × 2	
Input	Output
3	**6**
5	**10**
6	**12**
8	16
9	18
13	**26**
16	32

4.

Rule: + 3	
Input	Output
1	4
4	7
6	9
7	**10**
9	12
12	**15**
16	19

5.

Rule: + 10	
Input	Output
3	13
8	18
16	26
35	**45**
57	67
58	**68**
72	82

6.

Rule: × 5	
Input	Output
2	10
4	20
5	25
7	**35**
9	45
11	**55**
12	60

TEST PREP ANSWERS
Page 143

1.

Rule: + 5	
Input	Output
0	**5**
1	**6**
2	**7**
3	**8**
4	9
5	**10**
6	11

2.

Rule: − 1	
Input	Output
5	4
7	6
15	14
37	**36**
69	68
56	**55**
85	84

3.

Rule: × 5	
Input	Output
2	**10**
3	**15**
6	**30**
7	35
9	45
12	**60**
15	75

4.

Rule: + 2	
Input	Output
3	5
5	7
6	8
7	**9**
12	14
15	**17**
16	18

5.

Rule: + 9	
Input	Output
4	13
7	16
23	32
5	**14**
8	17
62	**71**
74	83

6.

Rule: + 11	
Input	Output
3	14
5	16
26	37
28	**39**
43	54
67	**78**
78	89

Name _____ Date _____

Formulas

The distance around a figure is called the perimeter.

262

Find the perimeter of each rectangle.

1.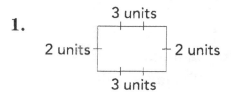

Perimeter = _____ units

Formula to find the perimeter of a rectangle:
P = 2 × l + 2 × w

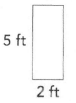

5 ft
2 ft

16 in.
25 in.

22 m
10 m

2. _____   3. _____   4. _____

Area is the number of square units needed to cover a figure.

263

Find the area of the rectangle.

5.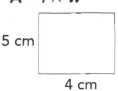

5 units
3 units

Area = _____ square units

Formula to find the area of a rectangle:
A = l × w

5 cm
4 cm

12 cm
16 cm

20 yd
30 yd

6. _____   7. _____   8. _____

Name _____ Date _____

Formulas

$$P = 2 \times l + 2 \times w$$
$$A = l \times w$$

Directions: Circle the letter of the correct answer.

1. Find the perimeter.   4 units   2 units

    A. 8 square units
    B. 6 units
    C. 12 square units
    D. 12 units

2. Find the perimeter.   3 ft   4 ft

    A. 7 square feet
    B. 14 feet
    C. 12 square feet
    D. 12 feet

3. Find the perimeter.   15 in.   20 in.

    A. 300 inches
    B. 70 inches
    C. 35 square inches
    D. 150 inches

4. Find the perimeter.   10 m   18 m

    A. 56 meters
    B. 180 meters
    C. 28 meters
    D. 360 square meters

5. Find the area.   4 units   3 units

    A. 12 units
    B. 12 square units
    C. 14 units
    D. 14 square units

6. Find the area.   6 m   2 m

    A. 8 square meters
    B. 8 meters
    C. 12 square meters
    D. 12 meters

7. Find the area.   11 cm   8 cm

    A. 19 cm
    B. 19 square cm
    C. 88 cm
    D. 88 square cm

8. Find the area.   7 yd   9 yd

    A. 63 yards
    B. 63 square yards
    C. 16 yards
    D. 16 square yards

PRACTICE ANSWERS
Page 145

1. 10
2. 14 ft
3. 82 in.
4. 64 m
5. 15
6. 20 square centimeters
7. 192 square centimeters
8. 600 square yards

TEST PREP ANSWERS
Page 146

1. D
2. B
3. B
4. A
5. B
6. C
7. D
8. B

Target Numbers

OBJECTIVES
- Write numerical expressions
- Use the rules for Order of Operations
- Practice whole number computation
- Use Guess, Check, and Revise

MATERIALS
- Calculators (optional)

TIME
- 40–50 minutes

TEACHER NOTES
- Introduce the activity using these examples.

  Add or subtract the given numbers to get a target number of 10. Use each number exactly once.

Numbers	Possible Answer
1 3 8	$3 + 8 - 1$
2 4 8	$8 + 4 - 2$

  Multiply, add, or subtract the given numbers to get a target number of 10. Use each number exactly once.

Numbers	Possible Answer
2 6 8	$2 \times 8 - 6$
1 2 5	$5 \times 2 \times 1$

- You may wish to have students use a calculator. Be sure students check first to see if their calculator uses the rules for Order of Operations.

EXTENSIONS
- Try this with the whole class or a small group. Use cards numbered 0 – 10. Decide on a target number. Mix up the cards and turn over 4 cards. See which student can make the target number first, using each number exactly one time and any operations. Numbers may be used as exponents and students may add parentheses.
- Use today's date. Use all but the last digit exactly once to get the last digit, the target number.

Example: 11/17/03

$1 + 1 + 1 + (7 \times 0) = 3$

Example: 4/28/02

$8 - (4 + 2) + 0 = 2$

ANSWERS
Answers will vary. Possible answers are given.

1. $2 + 4 + 3 + 1$
2. $5 \times 2 + 3 - 3$
3. $8 + 4 - 2 \times 1$
4. $5^2 \div 5 + 5$
5. $6 \div 3 \times 8 - 6$
6. $8 + 8 - 6 \times 1$
7. $6 + 4 \times 4 \div 4$
8. $4 \times 4 - 3 \times 2$
9. $8 + 4 \times 5 \div 5$
10. $4 + 4 + 4 \times 1$
11. $9 \times 2 - 2 \times 3$
12. $3 \times 5 - 1 - 2$
13. $8 \times 3 \times (8 - 7)$
14. $(9 - 1) \times (2 + 1)$
15. $(6 - 5 + 7) \times 3$
16. $4 \times 3 + 9 + 3$

Name _____ Date _____

Basic Operations
pages 32-99

Computing with
Whole Numbers
and Decimals
pages 144-207

Exponents
pages 98-99

Writing Expressions
page 251

Order of Operations
pages 253-254

Order of Operations
on the Calculator
page 418

Guess, Check,
and Revise
pages 376-377

Target Numbers

Write an expression equal in value to the target number.
- Use each of the four numbers exactly once.
- Use any operations.
- You may use a number as an exponent.
- You may use parentheses.
- You may change the order of the numbers.
- Use the rules for Order of Operations.

Target

10

1. 1 2 3 4 _____

2. 5 2 3 3 _____

3. 1 2 4 8 _____

4. 2 5 5 5 _____

5. 3 6 6 8 _____

6. 8 8 6 1 _____

7. 6 4 4 4 _____

8. 4 4 3 2 _____

Target
12

9. 8 4 5 5 _____

10. 1 4 4 4 _____

11. 2 2 3 9 _____

12. 1 2 3 5 _____

Target

24

13. 8 8 7 3 _____

14. 9 2 1 1 _____

15. 5 6 7 3 _____

16. 4 9 3 3 _____

Name _____ Date _____

Collecting and Organizing Data

266-267 A candy company gave 3 different candies to people to see which one they liked best. The person taking the survey kept track of each person's choice by making a tally mark.

1. Complete the chart to show how many tally marks were made for each candy.

Candy People Said They Liked Best

Candy	Tally	Number of People
Candy A		6
Candy B		13
Candy C		9

268 A video rental store keeps track of the kinds of videos people rent. Use the table to answer the questions.

Friday's Video Rentals

Type of Movie	DVD	VCR
Comedy	26	74
Family	18	87
Adventure	39	36

2. Were more DVD or VCR adventure movies rented? _____

3. How many VCR comedy movies were rented? _____

4. How many DVD family movies were rented? _____

5. How many VCR family movies were rented? _____

Name _____ Date _____

269

The ring on the left shows the children who bought popcorn and
the ring on the right shows the children who bought ice cream.

Snacks Children Bought

Popcorn   Ice Cream

Danielle   Jack   Michelle

Nikki   Paul

                  Pat

Natalie   Christine

6. Which children bought popcorn? _____

_____

7. Which children bought *only* ice cream? _____

_____

8. Which children bought *both* popcorn *and* ice cream? _____

_____

270–271

9. Yesterday Jill slept 8 hours, Vince slept 10 hours, and baby Therese
   slept 16 hours. Complete the pictograph to show this information.

Yesterday's Sleep

Name of Child	Number of Hours

Key [   ] = 2 hours of sleep

Name _____ Date _____

273–275   **10.** Ms. DiMillo asked the children in her class to choose which of three activities they liked best. Thirteen chose bicycling, 7 chose rollerblading, and 4 chose jogging. Make a bar graph to show this information. Be sure to write a title for the graph and label the horizontal axis and the vertical axis.

_____

15	
10	
5	
0	

Bicycling   Rollerblading   Jogging

_____

278–279   The stem-and-leaf plot shows the heights of players on the Del Mar Sharks soccer team in inches.

Heights of Del Mar Sharks

```
3 | 7 9 9
4 | 5 6 6 7 8 8 9
5 | 1 3 4 5
```

Key:
3l7 means a height of 37 inches.

11. How many players are 39 inches tall? _____

12. What is the height of the tallest player? _____

13. How many players are taller than 44 inches? _____

Name _____ Date _____

14. Dorothy measured her plant at the end of each week.
Here are the heights: Week 1: 2 inches, Week 2: 5 inches,
Week 3: 6 inches, Week 4: 11 inches, Week 5: 16 inches,
Week 6: 19 inches, Week 7: 22 inches, Week 8: 24 inches.
Complete this line graph to show how Dorothy's plant grew.

280–281

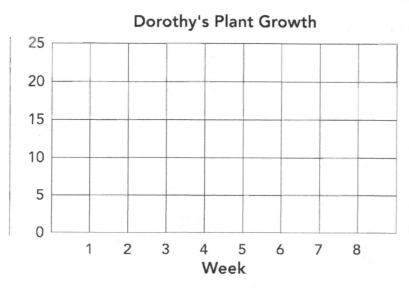

15. Dee Dee made a list of how she spent her time Monday.
Complete the circle graph of Dee Dee's day

282–283

Number of hours	Activity
10	sleep
6	school
3	play
5	other

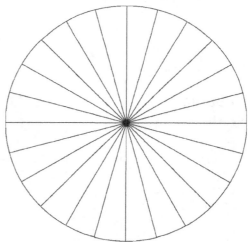

Name _____ Date _____

Collecting and Organizing Data

Graph the data in each table. For each, choose a format
from the choices on the right. Make the graph as complete
as you can with title, labels, and numbers.

1. Favorite Hobbies of Girls in Grade 4

Music	12
Sports	16
Art	8
Collecting	6

A. _____

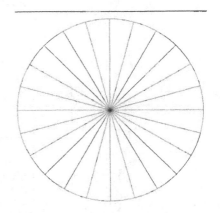

2. Temperature at School

8 A.M.	60°F
10 A.M.	65°F
12 noon	75°F
2 P.M.	80°F

B. _____

3. Favorite Subject of Students in Class

Math	8
Reading	6
Social Studies	6
English	4

C. _____

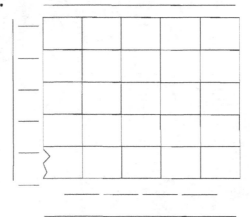

PRACTICE ANSWERS
Page 150

1. **Candy Children Said They Liked Best**

Candy	Tally
Candy A	ⵑⵑⵑ /
Candy B	ⵑⵑⵑ ⵑⵑⵑ ///
Candy C	ⵑⵑⵑ ////

2. DVD

3. 74 movies

4. 18 movies

5. 87 movies

Page 151

6. Danielle, Nikki, Natalie, Jack, and Pat

7. Michelle, Paul, and Christine

8. Jack and Pat

9. **Yesterday's Sleep**

Name of Child	Number of Hours
Jill	⬜⬜⬜⬜
Vince	⬜⬜⬜⬜⬜
Theresa	⬜⬜⬜⬜⬜⬜⬜

Key ⬜ = 2 hours of sleep

Page 152

10.
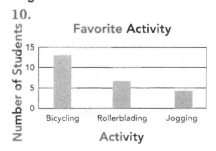
Favorite Activity

11. 2 players

12. 55 inches

13. 11 players

Page 153

14.
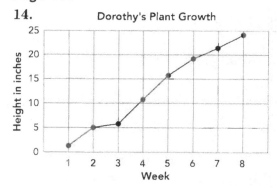
Dorothy's Plant Growth

15.
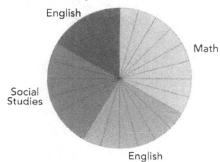
How I Spent My Day

TEST PREP ANSWERS
Page 154

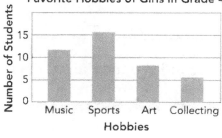
Favorite Subject of Students in Class

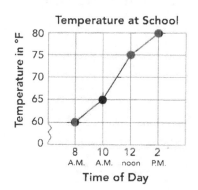
Favorite Hobbies of Girls in Grade 4

Temperature at School

Name _____ Date _____

Summarizing Data

284–285

1. What is the mean number

   of blocks in a stack? _____

2. These are the ages of four children playing soccer:
   7, 8, 9, 8. What is the mean age of the children?   _____

3. Stephane earned money walking
   people's dogs. What is the mean
   amount of money Stephane earned

   in a day? _____

Day	Money Earned
Monday	$2
Tuesday	$5
Wednesday	$4
Thursday	$3
Friday	$1

4. The prices of a large cheese pizza at three different stores are:
   $8, $12, and $7. What is the mean price of a large cheese pizza?

   _____

286–287 Find the median of each set of numbers.

5. 3, 7, 8 _____   6. 4, 12, 5, 9, 3 _____   7. 5, 6, 3, 8 _____

8. The Boys Under 10 soccer team
   ordered the following soccer shirts.

Player	Shirt Size
Trevor	8
Joseph	12
Trang	10
Raul	10
Miguel	8
Andrew	12
Ethan	8

   What is the median shirt size? _____

Name _____ Date _____

Find the mode of each set of numbers.

9. 2, 6, 8, 4, 6 _____   **10.** 6, 3, 7, 3, 7 _____   **11.** 4, 8, 5, 1 _____

12. What is the mode of the data below? _____

Heights of U.S. Soccer Players

Name	Height in inches
Michelle Akers	70
Susan Bush	64
Brandi Chastain	67
Mia Hamm	65
Jena Kluegal	63
Shannon MacMillan	65
Aly Wagner	65

Source: www.womensoccer.com/refs/usteamsch.html

Find the range of each set of numbers.

13. 45, 47, 49, 43 _____   **14.** 76, 73, 71, 68 _____

15. What was the range of temperatures in Dallas

between May 8 and May 11? _____

Dallas Weather, May 8-11

	Mon.	Tues.	Wed.	Thurs.
Highs	90°	90°	92°	92°
Lows	70°	72°	72°	72°

Source: www.msnbc.com/news/wea_front

Name _____ Date _____

Summarizing Data

Directions: Fill in the circle next to the letter of the correct answer.

1. The chart shows how far the Dunn family rode on their 3-day bicycle ride. What is the mean number of miles the Dunns rode each day?

Day	Miles
1	10
2	7
3	7

○ **A.** 6 miles   ○ **B.** 7 miles

○ **C.** 8 miles   ○ **D.** 9 miles

2. The chart shows last week's dog adoptions from a pet shelter.

Dog Adoptions

Day	Number
Monday	3
Tuesday	6
Wednesday	5
Thursday	2
Friday	3
Saturday	10
Sunday	12

What is the median number of dogs adopted per day last week?

○ **A.** 2 dogs   ○ **B.** 3 dogs

○ **C.** 4 dogs   ○ **D.** 5 dogs

3. Samantha bought one T-shirt for $6, one for $5, and one for $4. What was the mean price?

○ **A.** $2   ○ **B.** $3

○ **C.** $4   ○ **D.** $5

The table shows the number of children who read the *Polar Express* during the first 5 months of school. Use the table to answer questions 4 and 5.

Polar Express Readers

Month	Number of Readers
September	2
October	1
November	4
December	14
January	4

Source: Polar Express, Chris Van Allsburg, Houghton Mifflin ©1985

4. What is the mode of the data?

○ **A.** 2 readers

○ **B.** 4 readers

○ **C.** 14 readers

○ **D.** 5 readers

5. What is the range of the data?

○ **A.** 13 readers

○ **B.** 10 readers

○ **C.** 12 readers

○ **D.** 1 reader

PRACTICE ANSWERS
Page 156
  1. 4 blocks
  2. 8
  3. $3
  4. $9
  5. 7
  6. 5
  7. 7
  8. size 10

Page 157
  9. 6
  10. 3 and 7
  11. There is no mode.
  12. 65 inches
  13. 6
  14. 8
  15. 22°F

TEST PREP ANSWERS
Page 158
  1. C
  2. D
  3. D
  4. B
  5. A

Name _____   Date _____

Probability

292–293 Describe the likelihood of picking a card with a circle from each box without looking. Use a word from the word bank.

1.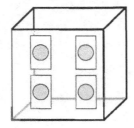

_____

2.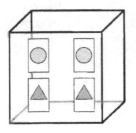

_____

3.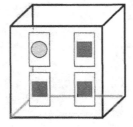

_____

4.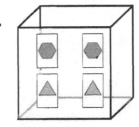

_____

5.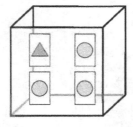

_____

6.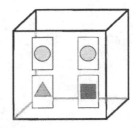

_____

294–295 In a number-cube game, the faces of the number cube are labeled with the numbers from 1 to 6. If you roll a 1, you win. If you roll any other number, you lose.

7. What are the possible outcomes when you roll the number cube?

_____

8. What is the probability that you win? _____

9. What is the probability that you lose? _____

Name _____ Date _____

10. If you flip a coin, how many possible outcomes are there? _____ 294–295

11. What is the probability of flipping a coin and getting tails? _____

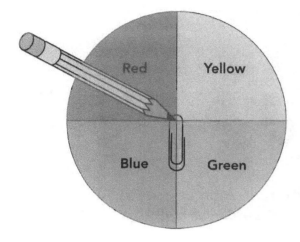

12. How many possible outcomes are on this spinner? _____

13. You want to spin either red or yellow. How many favorable outcomes are there? _____

14. What is the probability of spinning either red or yellow? _____

15. If you flip a coin 40 times, predict how many times you will get heads. _____ 297

16. When you flip a coin, what is the probability of getting heads?

_____

17. Chris has 3 different pairs of shorts and 3 different colors of shirts. List the different combinations of shirts and shorts Chris can wear. 298–299

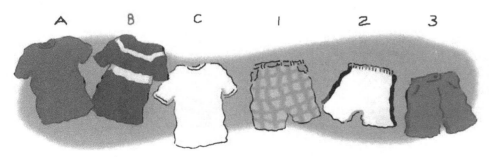

How many combinations are there? _____

Name _____ Date _____

Probability

For questions 1 and 2, write *certain*, *likely*, or *impossible*.

1. What is the likelihood of picking
   a black marble out of the bag

   without looking? _____

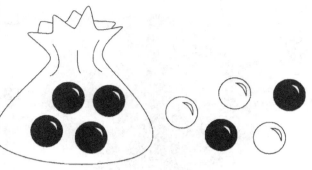

2. What is the likelihood of picking
   a white ball out of the bag

   without looking? _____

3. Each face of a number cube has a different number, from 1 to 6.
   When you roll the cube, what is the probability of rolling a 3?

   _____

4. If you have 3 boys and 3 girls, how many different ways
   can you pair them up, so that each girl has a boy partner?
   Show your work using drawings, words, or numbers.

   _____

PRACTICE ANSWERS
Page 160

1. certain
2. equally likely
3. unlikely
4. impossible
5. likely
6. equally likely
7. 1, 2, 3, 4, 5, 6
8. $\frac{1}{6}$
9. $\frac{5}{6}$

Page 161

10. 2
11. $\frac{1}{2}$
12. 4
13. 2
14. $\frac{2}{4}$ or $\frac{1}{2}$
15. Probability would suggest 20 times, although actual results may vary, just as student answers may vary.
16. $\frac{1}{2}$
17. A1   A2   A3
    B1   B2   B3
    C1   C2   C3

    9   combinations

TEST PREP ANSWERS
Page 162

1. certain
2. impossible
3. $\frac{1}{6}$
4. 9 different ways
   Check students' work.

Graph It!

OBJECTIVES
- Collect data
- Choose a type of graph to display data
- Summarize data
- Find the mean, median, and mode of a set of data

MATERIALS
- graph paper
- colored markers

TIME
- 50–60 minutes

TEACHER NOTES
- Brainstorm possible questions students can use for a survey.

- Review types of graphs that students might use to display their data.

- Review ways to find the mean, median, and mode.

EXTENSIONS
- Discuss the appropriateness of each type of graph used.

- Ask children to display their data in a second type of graph. Discuss which graph conveys the data most clearly.

ANSWERS
Answers will vary.

Use Assessment Rubrics on pages 8 and 9 to evaluate student work.

Name _____ Date _____

Collecting and Organizing Data
pages 266-283

Summarizing Data
pages 284-290

Graph It!

1. Take a survey of your classmates.

Here are some questions. Choose one of these or make up one of your own.

- Do you have a library card?
- At what age does a person become grown up?
- What time did you go to bed last night?
- How many letters are in your first name?
- Can you curl your tongue?
- How many doors does your house have?
- Do you think that the U.S. flag has more:
  a. long red stripes
  b. long white stripes
  c. short red stripes
  d. short white stripes?

Use a tally chart to keep track of the information as you collect it.

2. On a separate piece of paper, make a graph to display the data you collected.

3. Write about your graph. What conclusions can you draw? You may want to include information such as the mean, median, and mode. Does one way of comparing your data give a better idea of the "typical" or average person you surveyed?

_____

_____

_____

_____

Name _____ Date _____

Points, Lines, and Angles

302–304   **You can name a line by using any two points on the line.**

Draw and label an example of each.

	Write	Say	Your drawing with labels
1.	$\overleftrightarrow{AB}$	line AB	
2.	horizontal $\overleftrightarrow{CD}$	horizontal line CD	
3.	vertical $\overleftrightarrow{EF}$	vertical line EF	

Tell whether the lines are intersecting (but not perpendicular), parallel, or perpendicular.

4.

5.

6.

_____   _____   _____

305   **Line segments and rays are each a part of a line.**
A line segment has two endpoints.
A ray has one endpoint. It goes on in one direction.

Write whether each figure is a line, line segment, or ray.

7.

8.

9.

_____   _____   _____

Name _____ Date _____

An angle is formed when two rays meet at an endpoint. The endpoint is called the vertex of the angle.

A right angle forms a square corner.

An acute angle has a measure less than a right angle.

An obtuse angle has a measure greater than a right angle.

Write whether each angle is right, acute, or obtuse.

10.

_____

11.

_____

12.

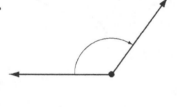

_____

A right angle is a 90° angle. A straight angle measures 180°.

Draw an angle that fits the description.

13. ∠ABC   **14.** 90° angle   **15.** straight angle

16. acute angle   **17.** obtuse angle   **18.** right angle

Name _____ Date _____

Points, Lines, and Angles

Write the name for each figure. Use the words from the word bank.

Word Bank		
vertical line	line segment	right angle
perpendicular lines	ray	obtuse angle
parallel lines	acute angle	180° angle

1.

_____

2.

_____

3.
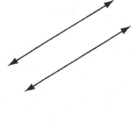

_____

4.

_____

5.

_____

6.
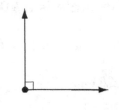

_____

7.

_____

8.
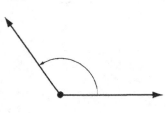

_____

9.
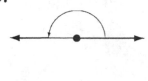

_____

PRACTICE ANSWERS
Page 166

Drawings for 1–3 may vary. Check students' drawings.

1.
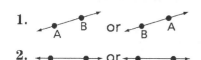

2.
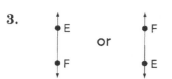

3.

or

4. perpendicular

5. parallel

6. intersecting (but not perpendicular)

7. ray

8. line segment

9. line

Page 167

10. acute

11. right

12. obtuse

Drawings for 13–18 may vary. Check students' drawings.

13.

14.

15.

16.

17.

18.

TEST PREP ANSWERS
Page 168

1. acute angle

2. vertical line

3. parallel lines

4. ray

5. perpendicular lines

6. right angle

7. line segment

8. obtuse angle

9. 180° angle

Name _____ Date _____

Plane Figures

310–311 **A polygon is a closed figure with sides that are all line segments.**

1. Which of these figures are polygons? _____

a.   b.   c.   d.

2. Write the number of sides for each polygon.

polygon	triangle	quadrilateral	pentagon	hexagon	octagon
Number of sides					

312–314 **3.** Which of these figures are not quadrilaterals? _____

a.   b.   c.   d.

Write the letter for the name of the figure.

4. _____   **a.** right triangle

5. _____   **b.** trapezoid

6. _____   **c.** obtuse triangle

7. _____   **d.** acute triangle

8. _____   **e.** rhombus

Name _____ Date _____

A rectangle is a parallelogram with four right angles.　313

9. Which of these figures are rectangles? _____

a.　　　　　b.　　　　　c.　　　　　d.

Triangles can be named by the lengths of their sides.　315

　　　equilateral triangle　　　isosceles triangle　　　scalene triangle

Write the name of each triangle.

10.　　　　　11.　　　　　12.

_____　　_____　　_____

Write the parts of the circle.　316

13. _____

14. _____

15. _____

16. If the length of the radius of a circle is 3 inches,
    what is the length of the diameter?　_____

Name _____   Date _____

Plane Figures

Directions: Fill in the circle next to the letter of the correct answer.

1. Which is a quadrilateral?

- ○ **A.**
- ○ **B.**
- ○ **C.**

2. Which is an acute triangle?

- ○ **A.**
- ○ **B.**
- ○ **C.**

3. How many sides does an octagon have?

- ○ **A.** 5
- ○ **B.** 8
- ○ **C.** 6

4. Which has a length double the length of the radius of a circle?

- ○ **A.** diameter
- ○ **B.** circumference
- ○ **C.** center

Name each polygon.

5.

_____

6.

_____

7.

_____

8.

_____

9.

_____

PRACTICE ANSWERS
Page 170

1. a, d
2. 3, 4, 5, 6, 8
3. a, b
4. b
5. e
6. d
7. a
8. c

Page 171

9. b, c
10. isosceles triangle
11. scalene triangle
12. equilateral triangle
13. radius
14. center
15. radius
16. 6 inches

TEST PREP ANSWERS
Page 172

1. C
2. A
3. B
4. A
5. square or rectangle or parallelogram or quadrilateral
6. hexagon
7. triangle, or equilateral triangle, or acute triangle
8. pentagon
9. quadrilateral or trapezoid

Name _____ Date _____

Working with Plane Figures

317   **1.** Which triangle is congruent to ? _____

a.   **b.**   **c.**   **d.**

318–319   **There are three ways to move a figure without changing its shape or size.**

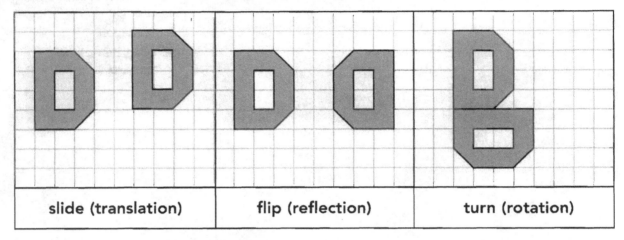

| slide (translation) | flip (reflection) | turn (rotation) |

2. Move the letter E on the grids below.

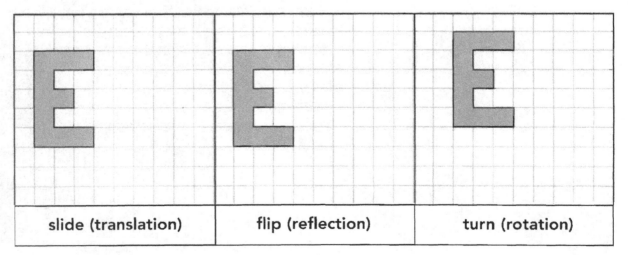

| slide (translation) | flip (reflection) | turn (rotation) |

Name _____ Date _____

3. Draw two similar triangles.

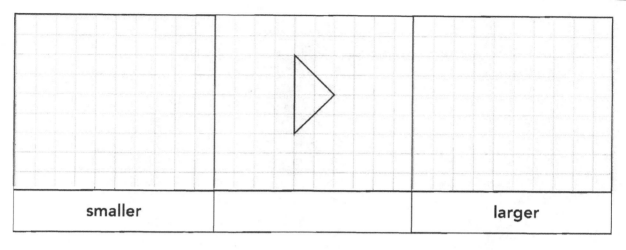

| smaller | | larger |

4. Use the grids. Draw two similar rectangles.

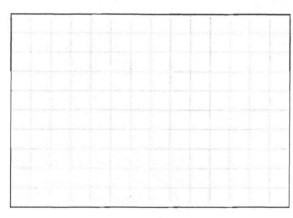 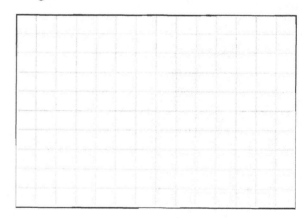

5. Use the grids. Draw two similar hexagons.

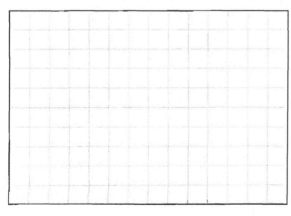 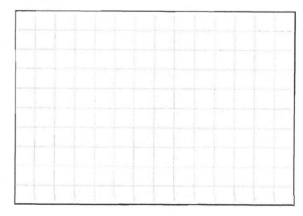

Name _____ Date _____

321   **6.** Use the grid to enlarge the picture. Copy one square at a time.

Name _____ Date _____

7. Circle the letters below that have line symmetry.
Draw the line or lines of symmetry.

A   B   C   D   E   F

G   H   I   J   K   L   M

N   O   P   Q   R   S

T   U   V   W   X   Y   Z

8. Choose one of the shapes at the side of the page.
Use the shape to make a tessellation.

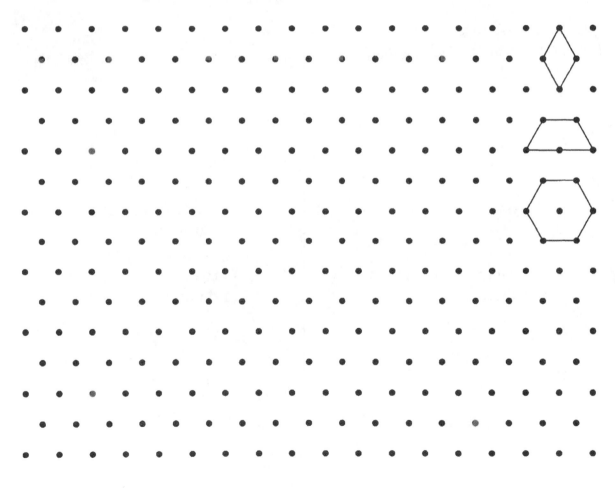

Name _____ Date _____

Working with Plane Figures

Directions: For questions 1–4, fill in the circle beside the answer to each question.

1. Which figure is congruent to this trapezoid?

○ **A.**

○ **B.**

○ **C.**

○ **D.**

2. Which picture shows a flip (reflection) of this figure?

○ **A.**

○ **B.**

○ **C.**

○ **D.**

3. Which figure is similar to this triangle?

○ **A.**

○ **B.**

○ **C.**

○ **D.**

4. Which figure shows a line of symmetry?

○ **A.**

○ **B.**

○ **C.**

○ **D.**

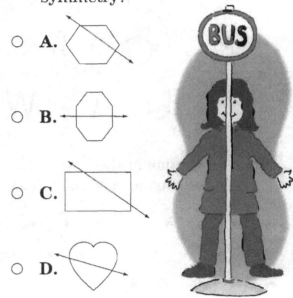

5. Use the back of this paper.

• Draw two figures that are similar.

• Draw two figures that are congruent.

• Explain the difference between similar figures and congruent figures.

PRACTICE ANSWERS
Page 174

1. b

2.

Drawings may vary. Check students' drawings.

Page 175

3.

Drawings may vary. Check students' drawings.

4.–5. Check students' drawings.

Page 176

6. Check students' drawings.

Page 177

7.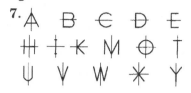

8. Check students' tessellations.

TEST PREP ANSWERS
Page 178

1. C

2. B

3. C

4. B

5. Check students' examples of similar figures and congruent figures.

Similar figures are the same shape, but can be different sizes. Congruent figures are the same shape and size.

Name _____ Date _____

Solid Figures

326–329

cube   square pyramid   rectangular prism   sphere

Complete the table.

	Solid figure	Number of faces	Number of edges	Number of vertices
1.	cube			
2.	square pyramid			
3.	rectangular prism			
4.	sphere			

Write the name for each solid figure. Use the word bank.

Word Bank				
sphere	rectangular prism	pyramid	cylinder	cone

5.

_____

6.

_____

7.

_____

8.

_____

9.

_____

Name _____ Date _____

Name the solid figure. 326-329

10. I have no faces, edges, or corners.

Who am I? _____

11. I have one curved surface and only have one face. Who am I?

(Remember: the face of a solid figure is flat.) _____

12. I have 6 square faces.

Who am I?_____

13. I have 2 faces that are circles.

Who am I? _____

14. I have 4 faces that are triangles.
I have 1 face that is square.

Who am I? _____

15. I have 4 faces that are long rectangles.
I have two faces that are square. Who am I? _____

Name the solid figure that you could make with each net. 330-331

16.

_____

17.

_____

18.

_____

19.

_____

Name _____ Date _____

Solid Figures

Write the name of each solid figure.

1.  _____

2.  _____

3.  _____

4. 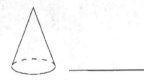 _____

5. _____

6.  _____

How many faces does each solid figure have?

7. rectangular prism _____

8. square pyramid _____

How many edges does each solid figure have?

9. cube _____

10. square pyramid _____

11. triangular pyramid _____

12. sphere _____

How many vertices does each solid figure have?

13. cone _____

14. square pyramid _____

15. cube _____

16. Use the grid to draw a net of a cube.

PRACTICE ANSWERS
Page 180

1. 6, 12, 8
2. 5, 8, 5
3. 6, 12, 8
4. 0, 0, 0
5. rectangular prism
6. cylinder
7. cone
8. sphere
9. pyramid

Page 181

10. sphere
11. cone
12. cube
13. cylinder
14. square pyramid
15. rectangular prism
16. cube
17. square pyramid
18. cylinder
19. triangular pyramid

TEST PREP ANSWERS
Page 182

1. cube
2. sphere
3. rectangular prism
4. cone
5. pyramid
6. cylinder
7. 6
8. 5
9. 12
10. 8
11. 6
12. 0
13. 1
14. 5
15. 8
16. Check students' nets to see if they can be folded to form a cube.

Pentominoes

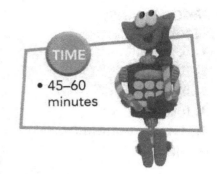

OBJECTIVES
- Make pentominoes.
- Use slides, flips, and turns.
- Find nets of open boxes.

MATERIALS
- Copies of 1-inch squared paper (page 224)

TIME
- 45–60 minutes

TEACHER NOTES
- You may wish to have students work in small groups.

- Have students sketch pentominoes and cut them out. Once the figures are cut out, students may slide, flip, or turn them to see if they have any duplicates.

EXTENSIONS
- Have students work in pairs. Each student makes a net that will fold into a cube and uses 5 colors to color the 6 squares so that no 2 faces next to each other are the same color. Each student then makes an exact duplicate of his or her net. Students then each take one of their nets and fold it into a cube. Next, students take turns observing their partner's cube from an angle so that only 3 of the faces are visible. Students try to use the matching net to predict which color is on the bottom of the cube.

- Have students try to find all possible nets for a cube.

ANSWERS

1. There are 12 different pentominoes.

2. The following 8 pentominoes can be folded into an open box.

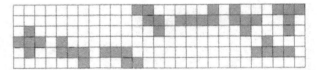

Name _____ Date _____

Pentominoes

Congruent Figures
page 317

Slides, Flips, and Turns
pages 318-319

Nets of Solid Figures
page 330

1. Use squared paper to see how many **pentominoes** you can make with 5 squares. Each square must share at least one full side with another square.

These are pentominoes.   These are not pentominoes.

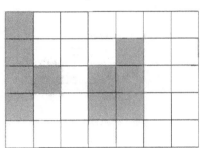 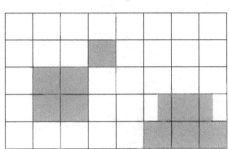

If one penomino can be slid, flipped, or turned to fit exactly on another, the two pentominoes are the same.

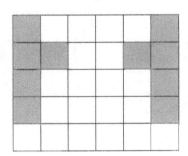

These pentominoes
are the same.

How many different pentominoes are there? _____

2. Predict which pentominoes can be folded into an open box. Check your guess by cutting and folding.

How many pentominoes can be folded into an

open box? _____

Name _____ Date _____

Time

334–336   Write the letter of the digital clock that shows the same time as the analog clock.

1. _____

    a. 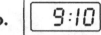 5:15

2.  _____

    b. 9:10

3. _____

    c. 10:30

4.  _____

    d. 10:00

5. _____

    e.  9:22

6.  _____

    f. 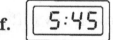 5:45

337   Write A.M. or P.M.

7. We saw the sun rise at 5:30 _____.

8. We watched the sun set at 7:15 _____.

Name _____ Date _____

9. How many seconds are in one minute? _____   `337-339`

10. The baby took a nap from 1:30 P.M. until 3:30 P.M.

How long was the baby's nap? _____

11. Soccer practice was from 3:30 P.M. until 5:00 P.M.

How long was soccer practice? _____

12. We drove from 10:45 A.M. until 11:23 A.M.

How many minutes did we drive? _____

13. The Eastern time zone is 3 hours later than the Pacific time zone.   `340`
If you wanted to make a call at 9:00 P.M. Eastern Standard Time
from the Pacific time zone, what time would you need to make

the call? _____

14. How many hours are in one day? _____ 2 days? _____   `341-343`

15. How many days are in one week? _____ 2 weeks? _____

16. Write the names of the months and the number of days in each month.

_____   _____   _____

_____   _____   _____

_____   _____   _____

_____   _____   _____

17. How many days are in a leap year? _____   `344`

18. How many weeks are in a year? _____

19. How many years are in a decade? _____ score? _____

century? _____ millennium? _____

Name _____　Date _____

Time

Directions: Circle the letter of the best answer.

1. What time is shown on the clock?

   A. 12:20

   B. 4:00

   C. 12:04

2. What time is shown on the clock?

   A. 9:06

   B. 6:08

   C. 8:30

3. What time is shown on the clock?

   A. 10:15

   B. 10:03

   C. 10:20

4. What time is shown on the clock?

   A. 2:47

   B. 2:57

   C. 2:42

5. The bus traveled from 8:30 A.M. until 10:50 A.M. How long did it travel?

   A. 2 hours

   B. 2 hours and 20 minutes

   C. 3 hours

6. Central Standard Time is one hour later than Mountain Standard Time. When you make a phone call from Denver at 6 P.M. (MST) to Chicago, what time is it in Chicago (CST)?

   A. 5 P.M.

   B. 6 P.M.

   C. 7 P.M.

7. How many days are there in 3 weeks?

   A. 3

   B. 21

   C. 14

8. What is the tenth month of the year?

   A. September

   B. December

   C. October

9. How many days are in February during a leap year?

   A. 28

   B. 29

   C. 30

PRACTICE ANSWERS
Page 186

1. d
2. c
3. a
4. f
5. b
6. e
7. A.M.
8. P.M.

Page 187

9. 60 seconds
10. 2 hours
11. $1\frac{1}{2}$ hours
12. 38 minutes
13. 6:00 P.M.
14. 24 hours; 48 hours
15. 7 days; 14 days
16. January, 31
    February, 28 or 29
    March, 31
    April, 30
    May, 31
    June, 30
    July, 31
    August, 31
    September, 30
    October, 31
    November, 30
    December, 31
17. 366 days
18. 52 weeks
19. decade: 10 years;
    score: 20 years;
    century: 100 years.
    millennium: 1000 years

TEST PREP ANSWERS
Page 188

1. B
2. C
3. A
4. A
5. B
6. C
7. B
8. C
9. B

Name _____ Date _____

Length

346 Write the customary unit that is about the size
of the benchmark indicated.

Customary Units of Length
inch
foot
yard
mile

1. length of a binder _____

2. width of a door _____

3. diameter of a quarter _____

4. four times around a football field _____

5. Find an item to measure using each unit of length.
Write each measurement.

Unit of measure	Item measured	Length
inch		
foot		
yard		

Complete.

6. 1 foot = _____ inches

7. 1 yard = _____ feet

8. 1 mile = _____ feet

The abbreviation for inch or inches is in. Write the abbreviation
for these units of measure.

9. foot or feet _____

10. mile or miles _____

11. yard or yards _____

Name _____ Date _____

Write the metric measure that is about the size of the benchmark indicated.

347

Metric Units of Length
meter
kilometer
centimeter
millimeter
decimeter

12. diameter of a crayon _____

13. height from the floor to doorknob _____

14. thickness of a dime _____

15. length of a new crayon _____

16. three times around a football field _____

17. Find an item to measure using each unit of length. Write each measurement.

Unit of measure	Item measured	Length
millimeter		
centimeter		
meter		

Complete.

18. 1 meter = _____ centimeters

19. 1 kilometer = _____ meters

20. 1 meter = _____ decimeters

The abbreviation for meter or meters is m. Write the abbreviation for these units of measure.

21. centimeter or centimeters _____

22. kilometer or kilometers _____

Name _____ Date _____

Length

Directions: Circle the letter next to the best unit for measuring each.

1. the width of a book

   A. yard

   B. mile

   C. inch

2. The height of a school building

   A. feet

   B. inch

   C. mile

3. The width of your finger

   A. meter

   B. centimeter

   C. kilometer

4. The distance a car drives in a week

   A. millimeter

   B. decimeter

   C. kilometer

5. the thickness of a piece of wire

   A. millimeter

   B. decimeter

   C. kilometer

Directions: Circle the letter next to the best estimate.

6. The height of this paper is about

   A. 24 inches

   B. 5 inches

   C. 11 inches

7. The width of this paper is about

   A. 4 cm

   B. 22 cm

   C. 88 cm

8. Your height is about

   A. 4 ft

   B. 14 ft

   C. 40 ft

9. The length of a new pencil is about

   A. 7 in.

   B. 7 ft

   C. 7 mi

10. The length of a new pencil is about

    A. 11 m

    B. 11 cm

    C. 11 mm

PRACTICE ANSWERS
Page 190
1. foot
2. yard
3. inch
4. mile
5. Answers will vary.
6. 12
7. 3
8. 5280
9. ft
10. mi
11. yd

Page 191
12. centimeter
13. meter
14. millimeter
15. decimeter
16. kilometer
17. Answers will vary.
18. 100
19. 1000
20. 10
21. cm
22. km

TEST PREP ANSWERS
Page 192
1. C
2. A
3. B
4. C
5. A
6. C
7. B
8. A
9. A
10. B

Name _____ Date _____

Perimeter, Area, and Volume

348–349 **Perimeter is the distance around a figure.**

Find the perimeter of each figure.

1.

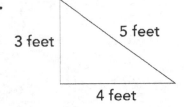

Perimeter: _____

2.

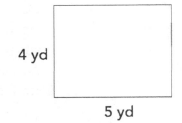

Perimeter: _____

3.

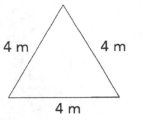

Perimeter: _____

4.

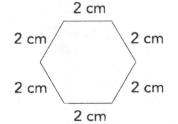

Perimeter: _____

5.

Perimeter: _____

6.

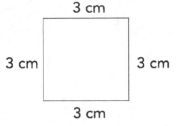

Perimeter: _____

7.

Perimeter: _____

8.

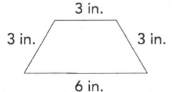

Perimeter: _____

Name _____ Date _____

Area is the number of square units needed to cover a figure.

350–351

Find the area of each figure.

9.

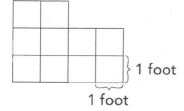

Area: _____

10.

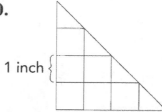

Area: _____

11.

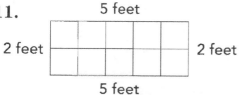

Area: _____

12.

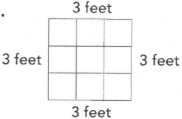

Area: _____

352–353

13.

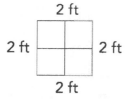

Area: _____

14.

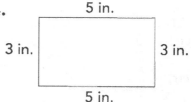

Area: _____

Find the perimeter and area of each figure.

348–353

15.

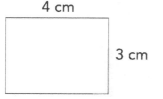

Perimeter: _____

Area: _____

16.

6 meters

4 meters

Perimeter: _____

Area: _____

Name _____   Date _____

Volume is the amount a solid figure holds.

354–355 Find the volume.

17. 2 units

1 unit

4 units

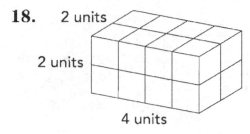

Volume: _____

18. 2 units

2 units

4 units

Volume: _____

19.

2 units

3 units

4 units

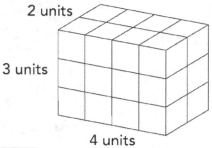

Volume: _____

Find the volume.

20. 2 cm

2 cm

2 cm

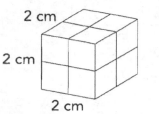

Volume: _____

21. 3 cm

2 cm

4 cm

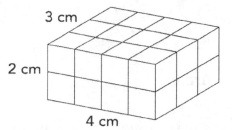

Volume: _____

22.

3 cm

3 cm

4 cm

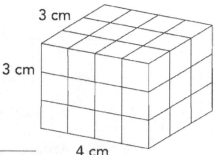

Volume: _____

Name _____ Date _____

Perimeter, Area, and Volume

Directions: Fill in the blank with a word from the word bank.

Word Bank
perimeter
area
volume

1. The number of square units needed to cover a figure

   is called the _____ _____.

2. You can use cubic units to measure the _____.

3. The distance around a figure is called the _____.

Directions: Circle the letter next to the correct answer.

4. What is the perimeter of the figure?

   3 cm
   2 cm

   A. 6 cm

   B. 10 cm

   C. 5 cm

   D. 12 cm

6. What is the area of the figure?

   5 ft
   3 ft

   A. 15 square ft

   B. 8 square ft

   C. 16 square ft

   D. 30 square ft

5. What is the area of the figure?

   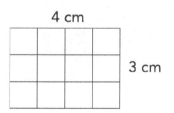

   4 cm
   3 cm

   A. 7 square cm

   B. 14 square cm

   C. 12 cm

   D. 12 square cm

7. What is the volume?

   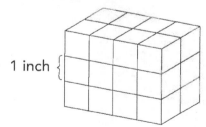

   1 inch

   A. 9 cubic inches

   B. 24 cubic inches

   C. 20 cubic inches

   D. 18 cubic inches

PRACTICE ANSWERS
Page 194
1. 12 feet
2. 18 yd
3. 20 in.
4. 12 cm
5. 12 m
6. 12 cm
7. 20 cm
8. 15 in.

Page 195
9. 10 square feet
10. 8 square inches
11. 10 square feet
12. 9 square feet
13. 4 square feet
14. 15 square inches
15. Perimeter:
   14 centimeters
   Area: 12 square centimeters
16. Perimeter: 20 meters
   Area 24 square meters

Page 196
17. 8 cubic inches
18. 16 cubic inches
19. 24 cubic inches
20. 8 cubic centimeters
21. 24 cubic centimeters
22. 36 cubic centimeters

TEST PREP ANSWERS
Page 197
1. area
2. volume
3. perimeter
4. B
5. D
6. A
7. B

Name _____ Date _____

Capacity

Write the unit that you would use to measure.

356

	Customary Units of Capacity
	teaspoon
	cup
	quart
	gallon

1. water in a swimming pool _____

2. cough medicine _____

3. glass of milk _____

4. bottle of milk _____

Circle a reasonable estimate for the capacity of each container.

5. soup bowl　　　　　2 cups　　　　　2 gallons

6. ice cream carton　　$\frac{1}{2}$ ounce　　$\frac{1}{2}$ gallon

7. drinking glass　　　8 pints　　　　8 fluid ounces

8. fish bowl　　　　　2 gallons　　　2 tablespoons

Complete.

9. 1 tablespoon = ____ teaspoons　10. 1 fluid ounce = ____ tablespoons

11. 1 cup = ____ fluid ounces　　12. 1 pint = ____ cups

13. 1 quart = ____ pints　　　　14. $\frac{1}{2}$ gallon = ____ quarts

Write the unit that you would use to measure.

357

	Metric Units of Capacity
	liter
	milliliter

15. water used to wash car _____

16. soupspoon _____

Circle a reasonable estimate for the capacity of each container.

17. bottle of juice　　2 milliliters　　2 liters

18. soupspoon　　　15 milliliters　　15 liters

19. milk carton　　　1 milliliter　　1 liter

Name _____ Date _____

Capacity

Directions: Circle the letter next to the best answer.

Which unit would you use to measure each?

1. spoonful of medicine

   A. teaspoon

   B. gallon

   C. quart

2. formula in baby bottle

   A. quart

   B. fluid ounce

   C. liter

3. pitcher of lemonade

   A. milliliter

   B. teaspoon

   C. liter

4. water in a bathtub

   A. gallon

   B. tablespoon

   C. cup

Which is a reasonable estimate for the capacity of each?

5. glass of juice

   A. 1 ounce

   B. 1 gallon

   C. 1 cup

6. can of paint

   A. 4 quarts

   B. 4 milliliters

   C. 4 teaspoons

7. injection of medicine at doctor's office

   A. 15 liters

   B. 15 milliliters

   C. 15 cups

8. student's bottle of liquid glue

   A. 8 fluid ounces

   B. 8 quarts

   C. 8 milliliters

PRACTICE ANSWERS
Page 199

1. gallon
2. teaspoon
3. cup
4. quart
5. 2 cups
6. $\frac{1}{2}$ gallon
7. 8 fluid ounces
8. 2 gallons
9. 3
10. 2
11. 8
12. 2
13. 2
14. 2
15. liter
16. milliliter
17. 2 liters
18. 15 milliliters
19. 1 liter

TEST PREP ANSWERS
Page 200

1. A
2. B
3. C
4. A
5. C
6. A
7. B
8. A

Name _____ Date _____

Weight and Mass

358   Write the unit that you would use to measure.

	Customary Units of Weight
1. weight of a handful of flour _____	ounce
2. weight of a truck _____	pound
3. weight of a turkey _____	ton

Circle the more reasonable estimate of the weight.

4. sack of potatoes   3 pounds   3 tons

5. raw hamburger   4 ounces   4 tons

6. dog   17 tons   17 pounds

7. elephant   2 tons   2 pounds

Complete.

8. 1 pound = ____ ounces   **9.** 1 ton = ____ pounds

359   Write the unit that you would use to measure.

	Customary Units of Mass
10. mass of a coin _____	gram
11. mass of an airplane _____	kilogram
12. mass of a bag of dog food _____	metric ton

Circle the more reasonable estimate of the mass.

13. sack of oranges   2 kilograms   2 grams

14. horse   1 metric ton   1 gram

15. handful of sugar   50 kilograms   50 grams

Name _____ Date _____

Weight and Mass

Directions: Circle the letter next to the unit you would use to measure each.

Directions: Circle the letter next to the most reasonable estimate of the weigh or mass of each.

1. weight of a hamster

   A. ounce

   B. kilogram

   C. metric ton

2. mass of a coin

   A. pound

   B. gram

   C. ton

3. weight of a dog

   A. ton

   B. gram

   C. pound

4. mass of a person

   A. kilogram

   B. gram

   C. metric ton

5. car

   A. 2000 pounds

   B. 2000 grams

   C. 2000 tons

6. newborn puppy

   A. 7 pounds

   B. 7 kilograms

   C. 7 ounces

7. truck

   A. 3 pounds

   B. 3 tons

   C. 3 kilograms

8. child

   A. 20 kilograms

   B. 20 grams

   C. 20 tons

PRACTICE ANSWERS
Page 202
1. ounce
2. ton
3. pound
4. 3 pounds
5. 4 ounces
6. 17 pounds
7. 2 tons
8. 16
9. 2000
10. gram
11. metric ton
12. kilogram
13. 2 kilograms
14. 1 metric ton
15. 50 grams

TEST PREP ANSWERS
Page 203
1. A
2. B
3. C
4. A
5. A
6. C
7. B
8. A

Name _____ Date _____

Temperature

Draw lines to connect the picture with the place on the thermometer where you would find that temperature.

A hot day
90°F

A warm day
75°F

Plain water freezes.
32°F

A snowy day
10°F

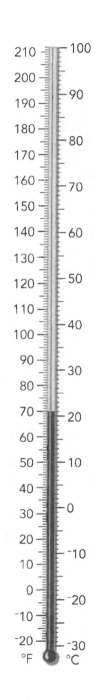

Normal body
temperature
37°C

Comfortable room
temperature
20°C

Plain water freezes.
0°C

Temperature inside
a freezer
−30°C

Math to Know

Name _____ Date _____

Temperature

Directions: Fill in the circle next to the letter with the best answer.

1. What temperature does the thermometer show?

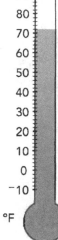

○ **A.** 76°F

○ **B.** 80°F

○ **C.** 72°F

○ **D.** 68°F

2. At what temperature does plain water freeze at sea level?

○ **A.** 0°F

○ **B.** 35°F

○ **C.** 32°F

○ **D.** 100°F

3. When you see the red liquid in a thermometer move upward, you know the temperature is getting _____?

○ **A.** cooler

○ **B.** snowy

○ **C.** close to freezing

○ **D.** warmer

4. What temperature does the thermometer show?

○ **A.** 10°C

○ **B.** 12°C

○ **C.** 15°C

○ **D.** 18°C

5. Water boils at one hundred degrees Celsius. How would you write the temperature for boiling water?

○ **A.** 100 >C

○ **B.** °C100

○ **C.** 100°C

○ **D.** C100°

6. Which of the following is a comfortable room temperature?

○ **A.** 20°C

○ **B.** 72°C

○ **C.** 0°C

○ **D.** 100°C

PRACTICE ANSWERS
Page 205

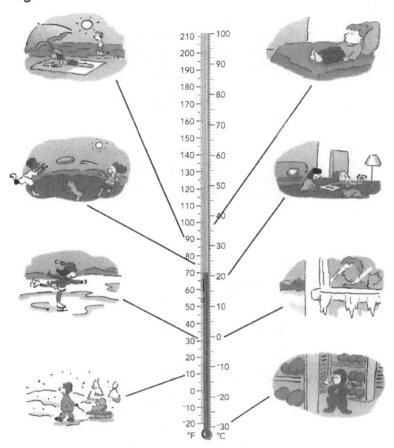

TEST PREP ANSWERS
Page 206

1. C
2. C
3. D
4. C
5. C
6. A

Name _____ Date _____

Computing with Measures

362–363

Customary Units of Length
1 foot (ft) = 12 inches (in.)
1 yard (yd) = 3 feet (ft)
1 mile (mi) = 5280 feet (ft)

Customary Units of Capacity
1 pint (pt) = 2 cups (c)
1 quart (qt) = 2 pints (pt)
1 gallon (gal) = 4 quarts(qt)

When you change from a larger unit, like feet, to a smaller unit, like inches, the number of units will be greater.

Complete.

1. 2 feet = _____ inches   2. 5 yd = _____ ft

3. 2 pints = _____ cups   4. 3 gal = _____ qt

When you change from a smaller unit to a larger unit, the number of units will be smaller.

Complete.

5. 24 inches = _____ feet   6. 18 ft = _____ yd

7. 32 cups = _____ pints   8. 24 qt = _____ gal

Complete. Think about whether you are changing from a larger unit to a smaller unit or from a smaller unit to a larger unit.

9. 2 yards = _____ feet   10. 2 mi = _____ ft

11. 9 feet = _____ yards   12. 15,840 ft = _____ mi

13. 16 pints = _____ cups   14. 10 c = _____ pt

15. 3 quarts = _____ pints   16. 2 pt = _____ c

Name _____ Date _____

Metric Units of Length	Metric Units of Mass
1 meter (m) = 100 centimeters (cm)	1 gram (g) = 1000 milligrams (mg)
1 meter (m) = 1000 millimeters (mm)	1 kilogram (kg) = 1000 grams (g)
1 kilometer (km) = 1000 meters (m)	1 metric ton = 1000 kilograms (kg)

When you change from a larger unit, like meters, to a smaller unit, like centimeters, the number of units will be greater.

Complete.

17. 3 meters = _____ centimeters   **18.** 5 m = _____ mm

19. 4 grams = _____ milligrams   **20.** 8 kg = _____ g

When you change from a smaller unit to a larger unit, the number of units will be smaller.

Complete.

21. 200 centimeters = _____ meters   **22.** 3000 m = _____ km

23. 5000 kilograms = _____ metric tons   **24.** 9000 kg = _____ g

Complete. Think about whether you are changing from a larger unit to a smaller unit or from a smaller unit to a larger unit.

25. 6 meters = _____ millimeters   **26.** 2 km = _____ m

27. 8000 meters = _____ millimeters   **28.** 8000 cm = _____ m

29. 60 grams = _____ milligrams   **30.** 2000 kg = _____ g

31. 3 metric tons = _____ kilograms   **32.** 2000 mg = _____ g

Name _____ Date _____

Computing with Measures

Directions: Solve each problem. Write your answer.

1. 1 liter is 1000 milliliters. How many milliliters

   are in 3 liters of juice? _____

2. 1 mile is 5280 ft. How many feet have you

   walked after $\frac{1}{2}$ mile? _____

3. You have 1 quart of milk. You have a recipe
   that asks for $1\frac{1}{2}$ cups of milk. Do you have
   enough milk to double the recipe?

   _____

4. Your skis are 4 feet long. You need to pack them
   in a box that is 40 inches long.

   Will your skis fit in the box? _____

5. A decade is 10 years. How many decades

   has a 50-year-old person lived? _____

6. 24 hours are in one day. How many days is 72 hours?

   _____

7. 7 days are in one week. How many days are in 4 weeks?

8. Soccer practice was from 3:30 p.m.
   until 5:00 p.m. How many minutes long

   was *soccer practice?* _____

PRACTICE ANSWERS
Page 208
1. 24
2. 15
3. 4
4. 12
5. 2
6. 6
7. 16
8. 6
9. 6
10. 10,560
11. 3
12. 3
13. 32
14. 5
15. 6
16. 4

Page 209
17. 300
18. 5000
19. 4000
20. 8000
21. 2
22. 3
23. 5
24. 9
25. 6000
26. 2000
27. 8,000,000
28. 80
29. 60,000
30. 2,000,000
31. 3000
32. 2

TEST PREP ANSWERS
Page 210
1. 3000 ml
2. 2640 ft
3. yes
4. no
5. 5 decades
6. 3 days
7. 28 days
8. 90 minutes

Treasure Hunt

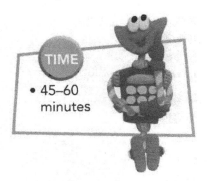

TIME
- 45–60 minutes

OBJECTIVES
- Measure using metric and customary units
- Estimate length measurements
- Estimate angle measurements

MATERIALS
- foot-long rulers, yardsticks, meter sticks
- small pieces of paper ($\frac{1}{4}$ sheet)
- treasure (This may be a nice note or drawing.)

TEACHER NOTES
- You may wish to have students work together in groups rather than individually.

- Help students use a ruler to devise a way to use paces (steps) to estimate distances in feet.

- Encourage students to write clues relating to mathematics. For example, use angle measurements and length measurements (both metric and customary).

- Be sure students have enough time to both make a treasure hunt and follow another student's treasure hunt.

EXTENSIONS
- Ask students to create a treasure hunt at home for a family member or neighborhood friend.

ANSWERS
Answers will vary.

Use Assessment Rubrics (pages 8 and 9) to evaluate students" work.

Name _____ Date _____

Treasure Hunt

HANDBOOK HELP

Customary Units
of Length
page 346

Metric Units
of Length
page 347

Measuring Angles
page 307

Plan a treasure hunt for a classmate.

1. Choose or make a treasure. Examples: a coin, a
    drawing, or a small toy.

2. Write Clue #1 on you first piece of paper. Include
    the exact starting point for your treasure hunt and
    directions for finding Clue #2. Be sure to use
    directions that involve mathematics.

Example:

> Clue #1:
>
> Stand with your back against the wall under the
> classroom clock. Move forward 10 feet. Clue #2 will
> be under the math book.

3. Write Clue #2. This will explain how to get from the
    location of Clue #2 to the location of Clue #3.

Example:

> Clue #2:
> Make a 90° turn to your left. Go 5 meters backward.
> Clue #3 will be behind the stapler.

Hide Clue #2 in the place that matches the directions
in Clue #1.

4. Write Clue #3 and Clue #4. Clue #4 will tell how to
    get to the location of the treasure. Hide the clues
    and the treasure in the places that match the clues.

5. Find a student to try your treasure hunt. Give the
    student Clue #1.

Name _____ Date _____

Problem-Solving Strategies

370–371   Use counters to act out this problem.

1. Suppose you put 24 cupcakes into groups. The first group has 3 cupcakes. Each group after that has 2 more cupcakes than the group before it. How many groups do you have? _____

372–373   Draw a picture to solve this problem.

2. You are sitting in the second row. Your best friend is sitting in the seventh row. How many rows are between you and your friend?

_____

374–375   Look for a pattern.

3. Predict how many squares will be in the 7th row. _____

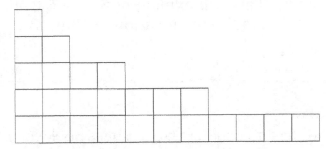

376–377   Guess, check, and revise to solve this problem.

4. There are 5 people. Some are riding bicycles and some are riding tricycles. There are 12 wheels.

How many bicycles are there? _____

Name _____ Date _____

Make a table to solve this problem.   378–379

5. You want to buy a computer game that costs $20.
You have $7 and are able to save $2 every week.
How many weeks will it take to have enough money? _____

Make a list to solve this problem.   380–381

6. You have 4 flavors of ice cream: vanilla, strawberry,
chocolate, and cookie dough. How many different
double scoop cones can you make that have 2 different
flavors? (Vanilla on the top and chocolate on the bottom
is not the same as chocolate on the top and vanilla

on the bottom.) _____

Work backward to solve this problem.   382–383

7. You need to meet your friends at the soccer field at 3 P.M.
It will take you 10 minutes to get to the soccer field.
You will need 20 minutes to get dressed and get your
soccer gear and water ready. At what time should you

start getting ready? _____

Write a number sentence to solve each problem.   384–385

8. You have 24 children in your class. You want to make
two cookies for each child. How many cookies will you

have to make? _____

9. You had $5 in your yesterday morning. Last night you
put $4 more in your wallet. This morning you bought
something that cost $7. How much money do you have

in your wallet now? _____

Name _____ Date _____

Problem-Solving Strategies

Directions: Use a problem-solving strategy to solve each problem. Show your work.

1. Suppose you cut a 9-inch stick into two pieces so that one piece is one inch longer than the other piece. How long is

   the longer piece? _____

3. Your dog needs 2 cups of dog food each day. How much dog food will you need for your dog for

   2 weeks? _____

2. 20¢ 15¢ 30¢

   pencil   eraser   notebook

   You bought 2 school items and spent 45¢. Which items did you buy?

   _____

   _____

4. How many squares are in the figure? (*Hint:* Count the number of of each size.)

   _____

PRACTICE ANSWERS
Page 214

1. 4 groups

2. 4 rows

X _X_ X X X X _X_

3. 22 squares

Row 1: 1 + 0 = 1

Row 2: 1 + 1 = 2

Row 3: 2 + 2 = 4

Row 4: 4 + 3 = 7

Row 5: 7 + 4 = 11

Row 6: 11 + 5 = 16

Row 7: 16 + 6 = 22

4. 3 bicycles

Bicycles	Wheels	Tricycles	Wheels	Total wheels	Results
1	2	4	12	14	Too high
4	8	1	3	11	Too low
3	**6**	**2**	**6**	**12**	**YES!**

Page 215

5. 7 weeks

Week	1	2	3	4	5	6	7
Amount saved	$7	$9	$11	$14	$16	$18	$20

6. 12 cones

v-s   s-v   c-v   cd-v

v-c   s-c   c-s   cd-s

v-cd   s-cd c-cd cd-c

7. 2:30

2:50 Leave for soccer field

2:30 Start getting ready

8. 48 cookies

$24 \times 2 =$ number of cookies to make

9. $2

$5 + $4 − $7 = money in wallet now

TEST PREP ANSWERS
Page 216

1. 5 inches

1	2	3	4	5	6	7	8	9

2. eraser and notebook

10 + 15 = 25 too low

20 + 30 = 50 too high

15 + 30 = 45 YES!

3. 28 cups or 14 pints or 7 quarts

Make a table.

no. of days	no. of cups
1	2
2	4
3	6
4	8
5	10
6	12
7	14
8	18
9	18
10	20
11	22
12	24
13	26
14	28

or

Write a number sentence.

$14 \times 2 = 28$

4. 14 squares

1×1 squares $\rightarrow 9$

2×2 squares $\rightarrow 4$

3×3 squares $\rightarrow 1$

Name _____ Date _____

Problem-Solving Skills

388–389

Which operation (addition, subtraction, multiplication, or division) would you use to solve the problem?

1. You have ■ cards and your friend has ▲ cards. How many more cards does your friend have? _____

2. You have ■ coupons. Your friend gives you ▲ more coupons. How many coupons do you have now? _____

3. You have ■ tickets that you share evenly among ▲ friends. How many tickets does each friend get? _____

390–391

Do you need an estimate or an exact answer? Find the estimate or exact answer.

4. You want to buy 3 pencils that cost 19¢ each. You have 50¢.

Do you have enough money? _____

estimate or exact? _____

5. You are buying a notebook. The clerk says you owe $3.18.

You give the clerk a $5 bill. How much change should you get? _____

estimate or exact? _____

392–393

Solve.

6. You buy 3 tickets for $1.50 each. You buy popcorn for $2 and a drink for $1.50. How much do you spend? _____

7. You buy a hamburger for $2 and a drink for $1.50. You give the clerk a $10 bill. How much change should you get back? _____

Name _____ Date _____

394-395

Solve.

8. My cat sleeps about 17 hours a day.
Does it sleep more than 1000 hours a week? _____

396-397

Use the table to solve the problem.

9. Andy, Barb, and Chris were in a race.
Andy is taller than the winner.
Chris finished 10 seconds after Andy.
In what order did the 3 children finish?

	1st	2nd	3rd
Andy			
Barb			
Chris			

1st _____ 2nd _____ 3rd _____

10. Diana, Emily, and Frank were in a
skating competition. Frank got
4 points fewer than Emily.
Emily is younger than the winner.
In what order did the 3 children rank?

	1st	2nd	3rd
Diana			
Emily			
Frank			

1st _____ 2nd _____ 3rd _____

398-399

Use the chart to solve.

Favorite School Lunch—Survey Results		
Lunch	Number of Girls	Number of Boys
Pizza	47	51
Hamburger	12	11
Peanut Butter and Jelly	10	6
Grilled Cheese	10	4
Hot Dog	2	3
Other	19	25

11. How many boys were in the survey? _____

12. How many girls were in the survey? _____

13. How many more students chose pizza than grilled cheese? _____

Name _____ Date _____

Problem-Solving Skills

Write the letter of the best answer.

1. Which operation would you use
   to find the number of mints in all? _____

   A. subtraction

   B. multiplication

   C. division

2. There are 24 students at 6 tables for a science lesson.
   Each table has 3 pounds of dirt. How much dirt is there?

   Which tells how to solve? _____

   A. There are 24 students at 6 tables, so you can divide.

   B. There are 3 pounds of dirt on each of the 6 tables,
   so you can multiply.

   C. There are 3 pounds of dirt for 24 students, so you can divide.

Solve each problem.

3. Chad tosses 2 beanbags. He scores 50 points.

   Where did his beanbags land? _____

   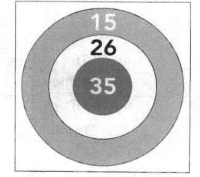

4. Stephanie tosses 2 beanbags.
   She gets more than 50 points. Find 3 different
   possible scores for Stephanie.

   _____ _____ _____

PRACTICE ANSWERS
Page 218
1. subtraction
2. addition
3. division
4. no, estimate
5. $1.82, exact
6. $8
7. $6.50

Page 219
8. no
9. 1st: Barb, 2nd: Andy 3rd: Chris

	1st	2nd	3rd
Andy	✗	✓	
Barb	✓		
Chris	✗	✗	✓

10. 1st: Diana, 2nd: Emily, 3rd: Frank

	1st	2nd	3rd
Diana	✓		
Emily	✗	✓	
Frank	✗	✗	✓

11. 100 boys
12. 100 girls
13. 84 more students

TEST PREP
Page 220
1. B
2. B
3. 15 and 35
4. 52, 61, 70

Field Trip

OBJECTIVES
- Compute with whole numbers and money
- Use problem-solving skills and strategies
- Use data from different sources to investigate a real-life situation

MATERIALS
- calculator (optional)

TEACHER NOTES
- Point out that *everyone* must go on the trip, even if it means ordering an extra bus that is not full.

- Students can do research or rely on their own experiences to solve problem 3.

EXTENSIONS
- Have students compare their estimated cost of a bought lunch to an estimated cost of a lunch brought from home.

- Have students work in groups to plan a class field trip researching all costs.

ANSWERS
You may wish to use the rubrics on pages 8 and 9 for evaluating students' work.

1. 3 buses

students	adults needed including teachers	Total	Buses Needed
116	23 + 1 = 24	140	3
+ 1 for the extra student		2 buses for 120 people and 1 extra bus for the 20 extra people	

2. $468

(116 students × $3 = $348;
24 adults × $5 = $120;
$348 + $120 = $468)

3. Answers will vary.

Name _____ Date _____

Field Trip

The 3rd and 4th grade classes at Solana Highlands School are taking a field trip.

Solve these problems. Explain your method using numbers, words, or pictures.

1. How many buses will be needed? _____

2. How much will the bus trip cost? _____

3. The students will eat lunch at a fast-food restaurant. If you were planning the trip, how much money would you ask each student to

   bring for lunch? _____

   How did you decide? _____

   _____

   _____

   _____

   _____

Handbook Help

Column Addition
  page 153

Dividing by a 1-digit Number
  pages 184-191

Interpreting Remainders
  pages 194-195

Dividing by Tens
  pages 198-199

Problem-Solving Strategies
  pages 369-385

Problem-Solving Skills
  pages 386-400

TEACHER	NUMBER OF STUDENTS
Ms. Doolittle	28
Mr. Summers	32
Ms. Petree	26
Ms. McCandless	30

ADULTS NEEDED
One adult is required for every five students. (Teachers count as adults.)

BUSES
Buses can hold 60 people. All adults and students must ride the bus.

BUS RATES:
Adults	$5
Children	$3

Name _____ Date _____